# Continuous Living

## In a Living Universe

by

David A. Ash

Published by:

Kima Global Publishers

50, Clovelly Road,

Clovelly

7975

South Africa

ISBN: 978-1-920535-77-3

eISBN: 978-1-920535-76-6

First edition August 2015

Second Edition February 2016

© Copyright David A. Ash 2015

Publisher's web site www.kimabooks.com

Author's web site: www.davidash.info

World rights Kima Global Publishers. All rights reserved. With the exception of small passages quoted for review purposes, no part of this publication may be reproduced, translated, stored in a retrieval system, or transmitted in any form or through any means including electronic, mechanical, photocopying, recording or otherwise without the permission of the publisher.

## Other books by David Ash

**The Vortex: Key to Future Science** (With Peter Hewitt) (Gateway)

**The New Science of the Spirit** (College of Psychic Science)

**Activation For Ascension,** (Kima Global Publishers 1995)

**The New Physics of Consciousness** (Kima Global Publishers 2007)

**The Role of Evil in Human Evolution** (Kima Global Publishers 2007)

**The Vortex Theory** (Kima Global Publishers 2015)

# Dedication

For my mother and my children in gratitude for their inputs, their love, their wisdom and their support.

*I write of Hell;*
*I sing and ever shall of Heaven,*
*And hope to have it after all*[1].

1    Robert Herrick 1591-1674, From an old Finnish Poem

# Table of Contents

| | | |
|---|---|---|
| Introduction | | 7 |
| Chapter 1 | Near Death Experiences | 9 |
| Chapter 2 | The Vortex Theory | 13 |
| Chapter 3 | An Encouter with Death | 17 |
| Chapter 4 | The Life Body | 19 |
| Chapter 5 | Extra Physical Reality | 26 |
| Chapter 6 | Genesis of the Life Body | 32 |
| Chapter 7 | A Reception from Hell | 37 |
| Chapter 8 | The Shadow Lands | 39 |
| Chapter 9 | The Attack | 41 |
| Chapter 10 | The Living Universe | 43 |
| Chapter 11 | The Hermetic Teaching | 49 |
| Chapter 12 | The Wipe Out | 52 |
| Chapter 13 | The Holographic Universe | 55 |
| Chapter 14 | The Evil God | 59 |
| Chapter 15 | Monotheism | 63 |
| Chapter 16 | The Dragon of Old | 67 |
| Chapter 17 | Karma | 71 |
| Chapter 18 | Saved by the Light | 74 |
| Chapter 19 | Forgiveness | 79 |
| Chapter 20 | Call the Midwife | 84 |
| Chapter 21 | My Experience of the Light | 88 |
| Chapter 22 | Plug into the Power | 95 |
| Chapter 23 | Earthbound | 99 |
| Chapter 24 | Journey into the Light | 101 |

| | | |
|---|---|---|
| Chapter 25 | God as Life | 103 |
| Chapter 26 | God as Energy | 106 |
| Chapter 27 | Faith or Fear | 110 |
| Chapter 28 | Intelligent Evolution | 115 |
| Chapter 29 | Personal Evolution | 123 |
| Chapter 30 | Taking Responsibility | 127 |
| Chapter 31 | Mirror Symmetry | 133 |
| Chapter 32 | The Polarities | 134 |
| Chapter 33 | Ascension and Healing | 138 |
| Chapter 34 | Understanding Ascension | 141 |
| Chapter 35 | The Greatest Opportunity | 143 |
| Chapter 36 | My Mythology | 147 |
| Chapter 37 | The End of an Age | 152 |
| Chapter 38 | The New Age | 161 |
| Chapter 39 | Conclusion | 165 |
| Afterword | | 167 |
| Appendix 1 | The Wisdom of Hermes | 171 |
| Appendix 2 | Mind the Matrix | 172 |
| Appendix 3 | Rebecca's Approach to Life | 173 |
| Appendix 4 | The Medicine Wheel of Life | 174 |
| Appendix 5 | A Positive Approach to Dying | 175 |
| Appendix 6 | Love in Action | 177 |
| Appendix 7 | Helping Others who are Earthbound | 178 |
| Appendix 8 | Happiness | 183 |
| Appendix 9 | Care for the Body | 185 |
| Appendix 10 | The Love Revolution | 186 |
| Appendix 11 | Self-Condemnation | 187 |
| Appendix 12 | Money | 189 |

## Chapter 1 – Table of Contents

| | | |
|---|---|---|
| Appendix 13 | The Universal Law of Love | 191 |
| Appendix 14 | The Native Elders | 192 |
| Appendix 15 | The Goddess | 193 |
| Appendix 16 | The Peace Education Program | 194 |
| Appendix 17 | Soul Fragments | 195 |
| Appendix 18 | The Dreamtime | 199 |
| Bibliography | | 200 |
| Index | | 203 |
| About the Author | | 206 |

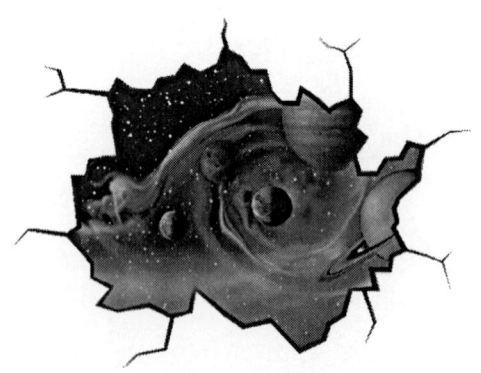

# Introduction

While developing *The Vortex Theory* it dawned on me that maybe the Universe is alive. Maybe it is a living body in which we are the equivalent of a colony of bacteria.

We get colonised by micro-organisms, including viruses, bacteria and parasites which can cause disease and even kill us. Our bodies have a protective immune system based on recognition or rejection. If a micro-organism is recognised it is allowed to live in the body. We all have colonies of bacteria that are tolerated in our guts. If an organism is not recognised it is rejected and the body will attempt to contain and destroy it. Our immune system has macrophages, cells that have evolved to engulf organisms that are not recognised.

If the Universe is alive maybe it has an immune system. Maybe it rejects organisms, destroying or containing them if they are not recognised. It struck me that if to the Earth we are as bacteria perhaps there is a Universal immunity that might reject us if we are not recognised as benign by the Universe. These were just speculations running in my head until a book was dropped in my lap. When I read it I was electrified. Its description of the neardeath experience of a man called Howard Storm suggested that my thoughts about Universal immunity might be dead right.

Howard Storm was a Professor of Art the University of North Kentucky. It was a weekend and he was in Paris when his duodenum perforated. The operation to save his life should have been performed within five hours but he was left for ten before surgery. He nearly died but when he recovered he spoke of a terrifying experience that shocks the complacent, spiritual and materialistic views of many people today.

Professor Storm was a confirmed atheist before he nearly died. When he recovered, his preconceived notions about death were shattered. He had expected oblivion but instead he had found himself in another body in which he was more alive than ever before. With a rigid belief in materialism and death he had been rejected and taken away by people of a similar disposition.

# Introduction

I then read of a New Zealander had no interest in spiritual matters. He didn't have strong beliefs about anything. He was a typical young surfer, traveling from wave to wave. When he nearly died, after being stung by deadly Box jellyfish, he found himself contained in a dark and dreadful space with other seemingly normal people. Were they rejected by the immune system of the Universe?

Attempting to use these near death experiences as evidence in support of a scientific line of enquiry is not easy. Firstly they are subjective and many attempts have been made to explain them away. I lean on a PhD thesis on near death experiences by Dr Penny Sartori and Dr. Sam Parnia on after death rescuscitations to overcome that objection. Secondly the two men involved became deeply religious after their experiences and their stories are couched in religious terminology. To overcome the objections religious language will obviously raise in many minds I have invented the word *relingo*. A relingo is a word in common usage charged with religious connotations. For example, the word *hell* is a relingo. Religious believers and disbelievers alike will react to words I cannot avoid using in my line of enquiry. My reference to a spiritual word as a relingo is not intended as a derogatory but to avoid its charge so its implications can be considered with a degree of impartiality.

Thirdly, although I am a scientist I am also deeply spiritual with my own mystical experiences and views on religion. I draw on these, as well as my science, in my attempt to make sense of these near death experiences and their implications for us all.

# Chapter 1

# Near Death Experiences

Because of the anti-spiritual prejudice in science, study of near death experience (NDE) is generally considered outside the domain of scientific enquiry. However, Dr Penny Sartori was awarded a PhD on her five year study of NDEs[1]. Her PhD demonstrated that the subject of near death experience can be included in science if it is approached it in a scientific manner. The achievement of the Dr Sartori thesis was to explain away every objection that has been raised by skeptics thereby establishing the possibility that NDEs could be real phenomena and included in a line of scientific enquiry.

Scientists may dismiss NDE evidence that human consciousness survives death as it seems to support religious belief. It is the idea that consciousness may survive death they don't like because of the obvious spiritual implications but many quantum physicists consider consciousness to be the bedrock of reality. It would be illogical for physicists to say consciousness cannot survive death if they perceive consciousness as a universal phenomenon. Max Planck was the father of quantum theory. He said, *"I regard consciousness as fundamental. I regard matter as a derivative of consciousness. We cannot get behind consciousness. Everything that we talk about, everything that we regard as existing, postulates consciousness."*

The most common dismiss of near death experience is that they are hallucinations in dying brains. If you are of that opinion consider the case of Eben Alexander. He is a neurosurgeon who

---

[1] **Sartori** P. The Near Death Experiences of Hospitalized Intensive Care Patients: A Five Year Clinical Study, Edwin Meller, 2008

## Chapter 1 – Near Death Experiences

had an NDE while in a coma for seven days with *E. coli* meningitis. He wrote,

*"During my coma my brain wasn't working improperly – it wasn't working at all…in my case the neocortex was out of the picture. I was encountering the reality of a world of consciousness that existed completely free of the limitations of my physical brain.*

*"As a practicing neurosurgeon with decades of research and hands-on work in the operating room behind me, I was in a better-than-average position to judge not only the reality but the implications of what happened to me.*

*"Those implications are tremendous beyond description. My experience showed me that the death of the body and the brain are not the end of consciousness; that human experience continues beyond the grave."*[2]

Dr. Alexander was a skeptic who believed in death before he had a life changing experience of continuous living. He now considers, from his own near death experience, that death is an illusion and human consciousness is not confined to the brain. He is not the only brain scientist frustrated by scientists who, because of the consensus belief in material existence and oblivion at death, limit consciousness to the brain. The neurophysiologist, Sir John Eccles wrote,

*"I maintain that the human mystery is incredibly demeaned by scientific reductionism, with its claim in promissory materialism to account eventually for all the spiritual world in terms of patterns of neuronal activity."*[3]

Scientific reductionism based on materialism is not true and just because it is the consensus view that doesn't make it true either! Science has proved everything is formed of energy and not material. Despite that, opinions that reality is limited to material are still hard wired in our brains.

---

2    **Alexander** E. *Proof of Heaven*, Piatkus, 2012
3    **Alexander** E. *The Map of Heaven*, Piatkus, 2014

## Continuous Living

Brain hardwiring occurs by the action of the neurochemical, *dopamine*. Dopamine sets up brain-cell circuits called *brain templates*. Some may argue that despite the fact that Storm and Alexander were of a materialistic mindset in adulthood they were exposed to Christian beliefs in America as children, when the brain templates most commonly form. However the vivid descriptions of their near death experiences didn't conform to what is normally taught in Sunday school.

Neither Storm nor Alexander was indoctrinated in a deeply religious family. Both were skeptic, materialists who didn't believe in life after death. They were typical of university trained academics. Before their NDEs, Storm was devoted to art and Alexander to science. Neither of them had time or inclination for other-worldly ideas before terminal illness struck, yet both had recoveries that defied medical science and both had NDEs that confounded their disbelief.

People who disbelieve are inclined to resist any challenge to their disbelief as surely as people who believe tend to defend their faith. People will often believe or disbelieve, regardless of evidence or reason, as opinions are rarely based on experience. People who go through a near death experience speak from experience not opinion. They believe in continuous living because they have had an experience of it.

NDEs are undeniably subjective but they do witness to the continuity of living beyond our world and they fill the pages of books and journals. I do not need to repeat the work of Dr Penny Sartori in validating them for scientific enquiry.[1] I have chosen to focus on NDEs because NDE descriptions fit with the line of enquiry I started when I realised we could be in a Living Universe; an idea that emerged as I worked on a new approach to physics outlined in *The Vortex Theory*[4].

My vortex theory is not conclusive and NDEs are subjective, but together they paint a picture that could revolutionise our

---

4   **Ash D**, *The Vortex Theory* Kima Global Publishing, 2015

## Chapter 1 – Near Death Experiences

understanding of the Universe and ourselves. This is especially true in the light of advances in medical science which enable people to be resuscitated after they have died.[5] How is it possible for a person to have a lucid out of body experience when they are clinically dead?

Limiting Life to biology based on carbon and water could be a very restricted view of reality. To expand our horizons on Life we could look toward physics rather than biology because with advances in physics the possibility of non-biological Life is emerging where carbon and water are not involved and even atoms are not required. To appreciate the possibility of a Living Universe it is necessary to rewrite physics. The inanimate view of the Universe held by classical physicists no longer holds true. Quantum physicists treat consciousness rather than material as the basis of being since energy, not material substance, has been discovered to underlie everything.

Physics is not an easy subject to grasp but I made a discovery that could render physics accessible to everyone. It is a way of understanding how energy forms matter that not only explains with ease what we have already discovered but points to new levels of discovery and frontiers for exploration. However, these are in territories traditionally held by religion. That raises considerable difficulty because of the conflict between religion and science. From my perspective religion is not rubbish because in my approach to science what is described as *spiritual* appears to be a real form of matter based on other levels of energy which rightfully belong in the realm of science not religion.

---

5    Parnia S. Erasing Death, Harper One, 2013

# Chapter 2

# The Vortex Theory

Dr Eben Alexander said we may listen to religious people but we believe in scientists. I approach the subject of continuous living predominantly through science rather than religion, because, over half a century, I have developed a scientific theory that supports the ideas of universal consciousness and other dimensions of reality.

But why would anyone believe my theory? The answer is it has enabled me to explain many properties of matter in terms of a single principle and it led me to predict the outcome of a future observation.[1] Stephen Hawking wrote,

*"A theory is a good theory if it satisfies two requirements: It must accurately describe a large class of observations on the basis of a modelthat contains only a few arbitrary elements, and it must make definite predictions about the results of future observations."*[2]

A prediction of the accelerating expansion of the Universe was in my vortex theory published in 1995[3]. Two years later, in 1997, observations of supernova explosions in distant galaxies were published which confirmed my prediction that the expansion of the Universe is speeding up.[4]

1 **Ash D**, *The Vortex Theory* Kima Global Publishing, 2015
2 **Hawking** Stephen, *A Brief History of Time* Bantam, 1987
3 **Ash D**, *The New Science of the Spirit,* College of Psychic Studies, 1995
4 **Perlmutter** Saul., et al. *Discovery of Supernova Explosions...* Berkeley National Laboratory, Dec. 16, 1997

## Chapter 2 – The Vortex Theory

The vortex theory is similar to string theory. String theories suggest the universe is formed of strings of vibration. I extend that idea to include spin. My theory is based on the idea that matter is formed of spinning energy; that subatomic particles are vortices of energy. That is why I call it *The Vortex Theory*. I got the vortex idea from ancient Yogic philosophy[5]. My approach to science is inclined to be more spiritual than material because Yoga is one of the most ancient spiritual traditions on the planet.

However, my idea of spirit is scientific, not religious, as I treat spirit as a non-atomic form of matter called *space plasma*. Non atomic matter may form electro-magnetic fields that could appear to be solid bodies. Existing in parallel dimensions, the fields could overlay bodies of atomic matter and mediate through them.

In my book, *The Vortex Theory* I draw on Einstein's theory of relativity to explain how this is possible. I also suggest that spirituality is based on attempts in ancient times to understand forms of matter we are only just beginning to appreciate with the rapid advance of science and technology.

When people have near death experiences they separate from their bodies. The separation is possible if consciousness mediates with the body through an electromagnetic field of life[6]. When the body dies, the field could live on. The mind would not be dependant on the brain if it functions in an electro-magnetic field that mediates through the brain[7]. When the brain dies, the mind-field could survive. This would explain near death experiences.

Before his NDE, Professor Storm was a cynic. He didn't entertain a remote possibility of continuous living. He believed death was

---

5 **Ramacharaka** Yogi, *An Advanced Course in Yogi Philosophy* 1904 (Cosimo facsimile 2007
6 **Burr** H. *Blueprint for Immortality*, Neville Spearman1972
7 **McTaggart** Lynne, *The Field*, Harper & Collins, 2001

# Continuous Living

the end of life but when he nearly died he found he was separated from his physical body in a seemingly identical body that continued to live on.

We may not believe in continuous living but if we do continue after so called death our existence would not be influenced by our disbelief. Storm did not believe in the continuity of life but discovering he was alive in another body all he could do was live on regardless. He had no choice in the matter. He couldn't suicide because the new body he was in didn't die even when it was attacked, torn apart and eaten alive. As Einstein said,

*"We have to accept things as they are, not how we think they should be."*

If we believe we have no future after death we may be tempted to live without consideration for the consequences of our beliefs and behaviour. The long term repercussion of this could be disastrous; as Howard Storm realised.

It is a matter of choice. We can choose to ignore developments in fringe science, such as the vortex theory, and the experiences of people who nearly or even actually die. We can choose consensus material belief and don the cloak of opinion offered by cynics to ward off the chill of things accepted in every human culture before the modern era, or we can choose to treat Storm's experience as a hell of a wake up call.

Applying science to spirit also makes sense of the relingoes heaven and hell because I believe if spirit is based on electro-magnetism, there would be a polarity to spirit because electromagnetism is polarized; positive and negative, north and south, on and off etc. In my opinion heaven and hell represent the polarity of spirit. Heaven and hell are depicted as poles apart. They are opposites. Without polarity and opposition there would be no opportunity to choose and without choice there would be no freedom. I believe we are here to choose between the polarities represented by heaven and hell.

In fact I think this is why we have choice; this is what human life is all about. I believe we experience both polarities on Earth to help us determine our destinations in continuous living. That is what Howard Storm discovered. In his NDE he was invited to go and live with people compatible with his attitudes. He was then

## Chapter 2 – The Vortex Theory

given a chance to choose otherwise. After that he came round in his physical body to testify to the consequences of his choices.

The very mention of heaven and hell produce a derisive knee jerk reaction in many people. Amongst many religious and spiritual people the idea of hell existing as a real place has been dismissed. Heaven and hell evoke pride and prejudice. To counter these reactions I present the account Howard Storm gave of his near death experience followed by that of Ian McCormack and leave you to decide for yourself what to make of them.

If you are a skeptic, that is a person who sees for him or herself rather than just believing on hearsay, I would recommend you read Dr Sartori for an overview of the validity of NDEs in general. She dismisses the hypoxia and hallucinatory hypotheses for NDE. Her book is intended for general readership[8] and her PhD thesis is available for in depth study[9].

Dr Eben Alexander's books are worth reading and Dr Sam Parnia's book Erasing Death[10], on after or actual death experiences, is difficult to explain away, even by a diehard debunker and there are many others available on this fascinating subject, not to mention blogs, websites, and interviews on the internet.

8 **Sartori P.**, The Wisdom of Near Death Experience, Watkins 2014
9 **Sartori** P. *The Near Death Experiences of Hospitalized Intensive Care Patients: A Five Year Clinical Study,* Edwin Meller, 2008
10 **Parnia S.** (with Josh Young) *Erasing Death,* Harper One, 2013

# Chapter 3

# An Encounter with Death

Howard Storm had struggled to stay alive in the hospital bed. An emergency operation was required to save his life but on Saturday the surgeons needed to perform it were not on duty in the Paris hospital. After five hours of agony, he surrendered to the inevitable. All he longed for was an end to the pain. Completely exhausted, he stopped fighting for breath closed his eyes and drifted into oblivion.

As recorded in his book *My Descent into Death*[1], Howard Storm was surprised to find he was standing up by his bed. The pain was gone and he felt altogether better. The floor was cool beneath his feet, the light was bright and everything in the room was crystal clear. Hospital smells assailed his nose. All his senses were heightened and alert. Anxious thoughts were racing through his mind but he felt more alive than ever before. He was awake. He was not dreaming. He spoke to his wife Beverly who was sitting by his bed but she didn't respond.

Then he noticed a lifeless body under the sheet and was surprised at the resemblance of its pallid face to his own. It couldn't have been him because he was looking down at it. Confused he shouted at his wife but she ignored him. He screamed and raged at her with mounting frustration but she just sat in her seat, drooped in utter despair.

Storm was used to shouting at his wife. Throughout his life he had been full of anger; anger against his father and the injustices of the world. He was angry with his kids and angry in his anxious determination to succeed. Achieving artistic acclaim and career ambition dominated his purpose. Storm was living the American dream with a philosophy that winner takes all.

1   **Storm** H, *My Descent into Death*, Clairview 2000

## Chapter 3 – An Encounter with Death

Compassion was not his thing and no thought was given to the purpose in life apart from the attainment of fame and financial security. In his opinion religion was a fantasy for the feeble minded and only uneducated, insecure people unable to cope with the realities of life believed in a continuity of life after death.

Storm turned to another patient and shouted directly in his face but the man looked straight through him as though he wasn't there. Then he looked back at the replica of himself on the bed. He was wondering why someone had made a wax model of him when the voices started. They came from outside in the hall. They were calling for him by his name in fluent English. That struck him as strange as the hospital staff were French.

# Chapter 4

# The Life Body

People may think death is the end but they may be wrong. Near death or an after death experience is when many people discover they have a second body that was with them all the time, but they didn't know about it. That was Howard Storm's dilemma. When he encountered death he discovered there was two of him and the awful reality he faced was the horrific consequence of his ignorance of this fact.

One of the most common features of near death experience is looking back at the body. It could be from the ceiling of an operating theatre or from a distance viewing the wreckage of a motor accident. Experiences of accentuated thought and feeling, with heightened faculties of sensation, are also spoken of by people recovered from near fatal accidents, operations or cardiac arrests, who came back from the brink of death.

NDEs usually report they are in a very real body in which they feel more alert and fully alive than ever before. People are confused when they look down at a lifeless body which they recognise as their own. They say they can see, hear and smell the scene around the body with greater clarity than before they were separated from it. People, for the most part, are not prepared for this.[1]

In the vast majority of NDEs people report of entering a tunnel of light, at the end of which is a being of light that fills them with reassurance and unconditional love. Beyond the tunnel is a whole new world similar to our own but full of light with a verdant landscape that is more lush and green and flowers with colours richer and brighter than on earth. There are trees there

1   **Sartori** P. The Wisdom of Near-Death Experience, Watkins, 2014

## Chapter 4 – The Life Body

and clear streams, mountains and a blue sky. There is no disease or suffering, but communities where people are cared for and places of learning and enlightenment.

In most reports of near death experience people speak of a paradise where they meet family members and friends deceased, very often grandparents, in clearly recognizable – albeit younger – bodies. All these people they have know to have died many years before; people whose bodies were either buried or cremated.

It is abundantly clear to anyone who has had an NDE that everyone who has ever gone before them has had two bodies too. The dilemma for those of us who have not had a near death experience is that it is hard to believe that we have two identical bodies that are engaged during physical life and disengage at the death of the physical body. It all boils down to the reliability of near death experience. Do we believe what people say when they have nearly died but have come back to life or do we discount their reports and attempt to explain them away.

Meanwhile, more near death and after death experiences occur, most of them going unreported. Those that are reported and are subsequently published in books or journals don't go away; they accumulate. The details may differ but the underlying pattern is the same either a tunnel of light leading to paradise or a fog leading to loss, confusion and darkness. The implication that death is not the end is undeniable and the two body dilemma is universal.

I believe NDEs and ADEs (Actual or After Death Experiences) are the most important things happening in the world today. If they are real everything else going on in the world pales into the insignificance of a dream because that is what NDEs report, compared to the world awaiting us after physical death, the world we live in now is tantamount to a dream.

If you are open to the possibility that NDEs are real, consider the two body dilemma. The experience of Howard Storm and many others suggests that we have an extra-physical body similar to the physical body we inhabit during our life of Earth that separates from it at death and continues in another dimension.

# Continuous Living

The faculties of life, like thought and sensation, seem more acute in the separated body – the normal one seems to dull them down – so it appears that life resides in the extra-physical body. For this reason I call it the *Life body* and suggest the physical body is a mantle covering and deadening it, so to speak. If this is so the expressions 'life-after-death' or 'afterlife' would be inaccurate. It could be we don't die at all but shed a physical mantle.

The body we live in on Earth could be likened to a space suit we occupy to visit a hostile planet. Death appears to be the equivalent of stepping out of the space suit when we go home. Many NDEs say the experience is like going home. It could be we are not the physical body we see but reside in a Life body we don't see.

Democritus

This information coming from near death episodes should be extraordinarily significant to science and yet it is sidelined because of speculative Greek philosophy and the opinion of one man in particular. One man's view of reality has had enormous impact, not just on the world view of scientists, but on the way most people perceive reality. People today who call themselves *atheist* or *skeptic* do so mostly because they have accepted the speculative philosophy of this individual from antiquity without question. Most people in the modern world live by it. Most men and women in science have blind faith in it. Some promote his outmoded philosophy with quasi-religious fervour! This man is Democritus, the father of materialism.

Science is responsible for uncovering truth but the tsunami of NDE evidence is ignored by most scientists because they follow the material philosophy of Democritus. Originator of the theory of the atom, Democritus lived in ancient Greece two and a half thousand years ago. He was vehemently opposed to the ideas of soul and spirit. His thinking influenced Aristotle who passed his anti-spiritual materialism onto the Western philosophy of science.

He denied the Egyptian perception of the delusion of death. Teaching only material atoms exist and everything else is opinion, he proposed the idea now dominating scientific

## Chapter 4 – The Life Body

opinion, that the observed world is the sole reality and death is the end of human life.

Plato's teaching of an immortal psyche (soul) acting as a template for the body has been overlooked. Scientists and people of a scientific bent follow the opinion of Democritus. People parrot his disbelief in soul and spirit and believe fervently in his atomic hypothesis instead. Modern society is based on his material philosophy. Supposedly educated people still adhere to moribund scientific materialism even when it was overturned a century ago by Albert Einstein.

Early in the 20th Century Einstein established that atoms are made of energy and that there is no material substance[2]. Einstein's groundbreaking equation $E=mc^2$ crashed the material hypothesis. Einstein established Planck's particle theory for energy we now know as Quantum theory[3]. Particles of energy are more particles of motion relative to the speed of light than particles of material moving at the speed of light. In my vortex theory I have been able to show particles of energy in vortex motion set up the appearance of material substance we experience as inertia and 3D extension.

Appreciating how spin in three dimensions forms the corpuscles we identify as subatomic particles, it is possible to see how energy sets up the illusion of material. If opposite directions of spin cause opposite electric charges then these properties of material can be 'explained away'. Accounting for mass and the forces of magnetism and gravity as vortex energy, the vortex theory shows how Democritus, deluded by spin, has misled humanity with his erroneous philosophy of materialism which could be described as *The Material Delusion!*

2   **Berkson**, William, Fields of Force: World Views from Faraday to Einstein, Rutledge & Kegan Paul 1974
3   **Kuhn**.Thomas, *Black-Body Theory and the Quantum Discontinuity: 1894-1912* Clarendon Press, Oxford, 1978

# Continuous Living

The vortex theory explains away the notion that we and our world are formed out of bits of material racing around in atoms. It shows how particles of energy are bits of movement that give rise to the illusion of material substance. In actuality there is nothing in the atom but activity. The particles of energy that make up atoms seem to be bits of activity in wave or vortex form. In reality, our world so seemingly solid and real appears to be just a bunch of impulses. Particles of energy are more like thoughts than things; the Universe to more like a mind than a material machine. To quote the renowned physicist Sir James Jeans,

*"Today there is a wide measure of agreement that the stream of knowledge is heading toward a non-mechanical reality; the universe begins to look more like a great thought than a great machine. Mind no longer appears as an accidental intruder into the realm of matter; we are beginning to suspect that we ought rather to hail it as the creator and governor of the realm of matter."*[4]

High energy physics has established that the world we live in is underpinned not by material but by motion we call energy. According to Einstein the most important thing about the particles of energy that make up matter and light is the speed. Einstein declared that the speed of light is the sole universal constant and everything in our world is relative to it. This law about energy can provide the key to understanding how we could have two bodies.

Howard Storm might have been prepared for the existence of his Life body if he had allowed for the possibility of parallel worlds. But Storm was a professor of art and he relied on scientific opinion which is opposed to any idea of unseen worlds and Einstein's rule that reality is constrained by the speed of light, would seem to support this view.

Particles of matter cannot be accelerated faster than the speed of light and energy with speeds faster than the speed of light

---

4     **Jeans** James, *The Mysterious Universe*, Cambridge. University Press, 1930

## Chapter 4 – The Life Body

cannot exist in the space-time of our world. That was established in Einstein's theory of relativity.

In the vortex theory, a particle of matter cannot be accelerated faster than the speed of light because it is driven by a quantum of energy which *is* a particle *of* the speed of light. In our world the energy *in* the vortex particle and the wave quantum are bits of movement *at* the speed of light. That is why everything in our world is relative *to* the speed of light.

Nothing can move faster than the speed of the energy forming it. That is obvious. Nonetheless it is theoretically possible for movement to exist *within* the vortex and quantum, with speeds faster than light; but not in our space-time. Energy with higher intrinsic speeds than the velocity of light would have to exist in other worlds.

Einstein's relativity may not allow energies beyond the speed of light to exist in the space-time continuum of our world but in the vortex theory space-time, as well as mass and the forces of electric charge, and magnetism, are the consequence of vortex motion. The vortex theory shows how one vortex of energy can set up the space and time through which another vortex moves. One vortex is the space-time continuum relative to which another is movement while waves of light wiz through them both.

The vortex theory allows for particles of energy to exist beyond the speed of light if they set up a space-time continuum relative to their faster speed of motion. Many other worlds, other planes of reality, could exist with their own space and time. With atoms of matter and photons of light, dependant on different constants of relativity to our own, these worlds could have bodies of matter appearing just like our own. The only difference would be the Einstein constant of relativity reflecting a faster speed of energy in the vortex particles and quantum waves that form them.

Science does not allow for the possibility of extra-physical worlds because science is devoted solely to the study of the physical world. But Professor Storm ended up in a dreadful situation because of this limit imposed on science by dogmatic faith in the material philosophy. It wasn't the delay in surgery that nearly led to the loss of Howard Storm; it was his lack of love and consideration for others and pursuit of the American Dream fueled by his belief in the philosophy of debunked scientific materialism.

Science is limited to the study of material things by scientism, an old defunct materialist pseudoscience. If material does not exist a science entirely devoted to the material to the exclusion of all other possible realities would be irrelevant. Science needs to drop the study of the premise of materialism if it is to remain relevant.

NDEs reveal that there is a continuity of life and consciousness beyond the apparent material. NDEs and ADEs are now part of scientific enquiry, despite the objections of disbelievers, because they number in many thousands and are impossible to ignore. People like Howard Storm could be having real experiences yet in the name of science NDEs and after death experiences are denounced as delusion, mostly by people who haven't experienced them or studied them properly. Taking into account the vortex theory, anyone with an open mind who takes time to study NDEs will realise the delusion lies with materialism, not the awkward facts pertaining to NDEs.

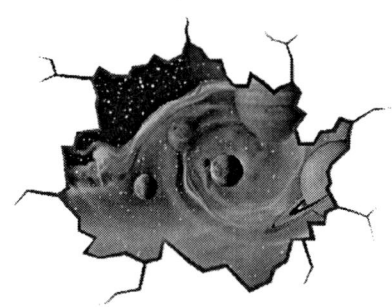

# Chapter 5

# Extra-physical reality

In my book *The Vortex Theory* I include a number of quotes from Richard Feynman. Feynman said *"The laws of inertia have no known origin"*[1]. Spin sets up inertia so the vortex of energy could provide an account for the inertia of matter. Feynman also said *"In nuclear energy we have the formulas for that but not the fundamental understanding."* The vortex shows how the dynamic state of energy can be stored in the static state of matter and nuclear energy could be stored energy escaping from the vortex when it unravels.

Feynman commented, *"It is important to understand in physics today we have no understanding what energy is"*. Energy is movement. Particles of energy are particles of activity at the speed of light. These occur as waves in light and vortices in matter. As one vortex extends and overlaps another the forces of interaction, including electric charge, magnetism and gravity, could be due to the inherent dynamic nature of the vortices.

When asked to put his theory of relativity in a single sentence Einstein replied, *"Remove matter from the Universe and you also remove space."*[2] If space is the infinite extension of vortex energy then space would be an extension of matter. Remove matter and obviously you would remove space.

Stephen Hawking said, *"I am hopeful we will find a consistent model that describes everything in the Universe. If we do that, it will be a real triumph for the human race."*

---

1     **Feynman** R. The Character of Physical Law, Penguin 1992
2     **Clerk** R.W. *Einstein: His Life & Times* Hodder & Stoughton 1973

# Continuous Living

The vortex theory, combined with wave theory, provides a consistent model to describe everything in the Universe. As well as accounting for the forces and particles of nature, and those discovered in high energy physics, the vortex and wave theory together provide a simpler way of comprehending quantum theory than quantum mechanics. As Richard Feynman said, *"Quantum mechanics…you never understand it, you just get used to it!"*

When the vortex is incorporated in wave theory a scientific model emerges simple enough for anyone to understand. It explains things that mainstream physics is unable to explain. Because of the success in predicting the accelerating expansion of the Universe, the vortex theory is as sound as any other theory in physics. You don't have to believe the standard model. You can believe the wave-vortex model instead because, as well as complying with the scientific method, it brings together the theories of relativity and quantum theory, which the standard model is unable to do; so it is complete.

In the concluding paragraph in his book *A Brief History of Time*, Stephen Hawking said, *"If we do discover a complete theory, it should in time be understandable in broad principle by everyone, not just a few scientists. Then we should all, philosophers, scientists, and just ordinary people be able to take part in the discussion of the question of why it is that we and the universe exist. If we find the answer to that, it would be the ultimate triumph of human reason - for then we would know the mind of God.*[3]

The evidence is in and an understandable complete theory to explain it has arrived. We have to choose whether to go along with it or retreat into cynical denial and bury our brains in quark theory and quantum mechanics. But they too have been overturned by the vortex theory! Quark theory is disproved by awkward facts about the proton and as for quantum mechanics that was disproved by the neutron in the 20th century but no one in physics dared admit it. There has been a big cover up in

3    **Hawking** Stephen, *A Brief History of Time*
     Bantam Press 1988

## Chapter 5 – Extra-physical reality

quantum physics! Science is not as honest and truthful as scientists like to make out.

You may prefer to disbelieve me but Eben Alexander, Howard Storm and the New Zealand surfer Ian McCormack were disbelievers too before they nearly died. Then during their NDEs they all had a profound experience of the reality of continuous living and they all returned to their physical bodies and recovered full health in total defiance of medical science.

They now believe in continuous living and miracles as well! No one survives a week of E. coli meningitis without severe brain damage, no one normally survives a perforated duodenal ulcer unless they receive emergency surgery within five hours and Ian McCormack was in the mortuary after succumbing to the stings of up to five separate Box jellyfish.[4] One is usually enough to kill anyone! All three men had after death experiences.

The evidence of NDEs and actual death experinces suggest we all have two bodies right now, one permeating the other, and we cannot die because our life and consciousness, our mind memories and emotions are seated in the one that doesn't perish; the one we continue to live in when we shed our mortal coil like a worn out coat.

People with a religious or spiritual bent believe in a soul which could correspond to the Life body. This could be linked in some way to *The Field*[5] and the *Life field* featured in books and scientific papers.

To understand the Life body and the way it relates to the physical first we have to accept that the particles of energy that make up matter are bits of movement. In the physical body the speed of movement is the speed of light. In the extra-physical Life body it would be some speed faster than that of light. They are both made of matter but neither of them have any material

---

4  **Sharkey,** J., *Clinically Dead* Amazon 2012
5  **McTaggart** Lynne, *The Field,* Harper & Collins, 2001

substance. We know that materialism is a myth and everything is energy.

Storm reported he saw more clearly than before. He smelt every thing in the room acutely. He said he had a strong taste in his mouth and could hear the blood coursing through the body he was standing in when he looked down at the lifeless body on the bed.

He emphasised in his book and the *Focus* television interview on the internet that he clenched his fists to see if he was awake and not dreaming. They felt real and solid! He said could still see and shout at his wife but why was she totally unaware of him?

Howard Storm was standing alive and well in what he perceived to be his body alongside his own corpse. He said he had all his faculties and was more alive than ever before. This out of body pheomenon is a common experience in near and actual death episodes.

It is common knowledge a car can go at the speed of a bike but there is no way a bike can keep up with a car when it accelerates. This shows that lesser speeds are a part of greater speeds but greater speeds are not a part of lesser speeds. I call this self evident law of motion the *Law of Speed Subsets*.

All energy at or below the speed of light is part of the physical plane relative to the speed of light. Energy above the speed of light would not be part of the physical plane. It would occur in its own space-time continuum relative to a higher energy speed. This would explain why, in an NDE lasting for only a few minutes or a few hours, the subject has the experience of being in another reality for ages. Out of the physical and in the extra-physical, the individual would be in a different expanse of space and a different flow of time. Years could pass there when they are out for only minutes or hours here, or visa versa.

Because the speed of light is a part of the higher speed of *super-energy*, physical space and time would be a *subset* or a part of the space-time continuum of the super-energy, extra-physical

## Chapter 5 – Extra-physical reality

plane of reality. This subset law explains why Howard Storm[6], in his Life body, was able to stand in the hospital room, look at his physical body, smell the hospital smells and feel the linoleum floor beneath his feet and the air on his skin. He could see and shout at his wife because she was in his world - physical space time being a subset of the extra-physical – but she couldn't see or hear him because the extra-physical isn't part of the physical.

Howard was standing right next to his wife Beverly. He could see her because the physical light reflecting off her was still a part of his reality. She could not see or feel him because he was not in her level of space and time so physical light would not reflect off his extra-physical Life body.

For her, the light was passing through that space so had she looked up she would have seen only the room behind him. As the sound vibrations from his voice were not in her level of space and time she could not hear him. His sound went right through her without any interaction with her body. For her there was nobody there, just a lifeless body on the bed. That was why she was so distressed. She thought her husband was dead. She didn't know he was standing there shouting at her. As Howard was not in Beverly's level of the space-time continuum he was invisible, inaudible and intangible to her.

It was a very frustrating situation for Howard. He could see his wife was in despair but he was unable to tell her he was OK. That must have been very upsetting. He was stressed and angry and took it out on the other patient in the room but his ranting and raging was to no avail as that other person in the physical world couldn't see or hear him either.

But others in Storm's new level in space and time were aware of him. Other people already in the extra-physical reality could see him and hear him and speak to him. They knew he was there. They were expecting him. They were calling out to him in his own name. He was lonely and afraid. Nothing made sense to him. Utterly unprepared and ignorant of his situation he had no

6    **Storm H.**, *My descent into Death*, Clairview 2000

# Continuous Living

idea of the peril he was in if he responded to the voices. Thinking they might be able to help him, he headed toward them. Ignoring a premonition of danger and against his own better judgment he left the safety of the room and entered the hall. Immediately he was shrouded in fog.

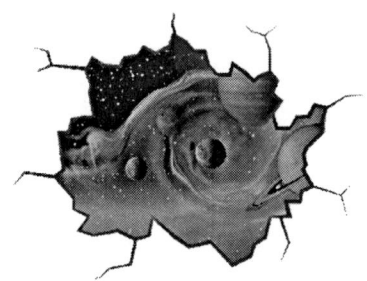

# Chapter 6

# Genesis of the Life Body

Howard Storm was vulnerable to the voices because of his mindset. One of the most common recalls of NDEs and ADEs is meeting grandparents after the transition we call death. Because Howard Storm closed his mind to the possibility of continuous living he didn't allow for the possibility of survival of his grandparents or any other friends or family gone before him. Because of his attitude of denial, by his own choice, he cut himself off from anyone who really cared for him. No one nice came to greet him.

If, in an NDE or ADE, someone recognises family members who departed before them, these people must have been in Life-bodies which grew alongside their physical bodies to be recognizable. We know the physical body of each individual is unique – apart from identical twins – because of the unique combination of parental genes at conception. This suggests that in its own level of space and time the extra-physical Life-body has grown in parallel to the physical. It seems to have been laid down in its own space and time alongside the apparent body in our world.

The research of Harold Saxton Burr[1] suggests an electromagnetic field, which he called the life field, may play a part in the differentiation of the physical body. Maybe it supports the differentiation of the Life-body too!

Burr was a professor of medicine at Yale University. His research findings are supported by animal stem cell research which shows that when tissues are regenerated from undifferentiated

1   **Burr** H.S, *Blueprint for Immortality*, Neville Spearman, 1972

## Continuous Living

stem cells, the cells organise themselves into the tissue as if responding to some hidden blueprint.

Burr discovered that an electromagnetic field overlies every living organism, acting as a blueprint, which influences its morphology. Burr's research appeared to confirm the Platonic view of the soul acting as a template for the body. But who has heard of Burr, have you? His research has been sidelined, ignored or dismissed because it conforms to the world view of Plato not his recalcitrant pupil Aristotle. In science as in religion it appears only those who conform to accepted beliefs are taken seriously. In his experimental research, for which he published over thirty papers, Burr realised that it is not what a cell is genetically that determines what it will become but where it is in respect to the life-field.

If the extra-physical Life body is a replica of the physical, because it forms with the physical, then Burr's differentiating field, playing a role in the formation of the physical body, might possibly influence the morphology of the extra-physical. How else could both bodies form in parallel?

There are a lot of unanswered questions but my contribution, from the vortex theory, is to provide a way of explaining how an energy field and a parallel body in a dimension based on a higher speed of energy could overlay a living organism in the physical dimension.

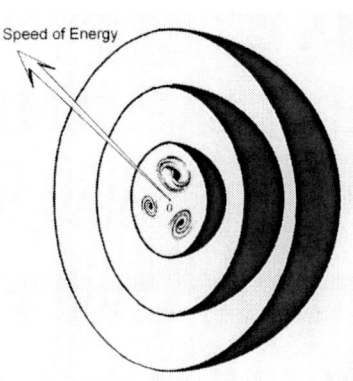

If vortices of energy establish the space time continuum in the physical world then vortices of extra-physical energy could set up a space-time continuum in the extra-physical world. The physical and extra-physical could then coincide in the same *here and now* because the separation between the physical and the extra-physical would not be space and time. The dimension separating the planes of physical and extra- physical energy would be speed; the intrinsic speeds of the energies.

## Chapter 6 – Genesis of the Life Body

For many years I have postulated a model for the Universe based on three main levels of energy which were then subdivided into further levels. The model is based on a system of concentric spheres; similar to the model I use for the subatomic vortex of energy to account for electric charge, magnetism, gravity and space-time.

The concentric spheres depict the planes of reality. They could also be thought of something like a set of nested Russian dolls. The centre of the set would be zero speed. The innermost sphere would represent the speed of light and depict the physical plane of reality. The second sphere would depict the extra-physical plane of reality. The third sphere in this set would represent a super-physical plane. This model corresponds to the Hermetic model of planes of reality represented by the *Harmony of the Spheres*.

The possibility of coincidence between physical and extra-physical realities explains how Howard Storm could have had an extra-physical Life body totally coincident with his physical body, during his time on Earth, so that the Life-body could slip out of the physical, like a snake shedding its skin, when his time in physical reality was up. It is as though the physical body is scaffolding for the extra-physical and death is just the scaffold falling away to reveal the real body.

In my first book, *The Vortex: Key to Future Science*[2] (with Peter Hewitt), I predicted a process called DNA resonance which would enable the frequencies in the extra-physical to mediate through the physical.

The DNA molecule is reminiscent of the coil in a radio or television set. The coil resonates with the broadcast signal transforming the programme it carries into electric impulses. The DNA molecules in all the trillions of cells in the physical body could be resonating with the DNA molecules in the

---

2  **Ash** D & **Hewitt** P *The Vortex: Key to Future Science,* Gateway Books 1990

## Continuous Living

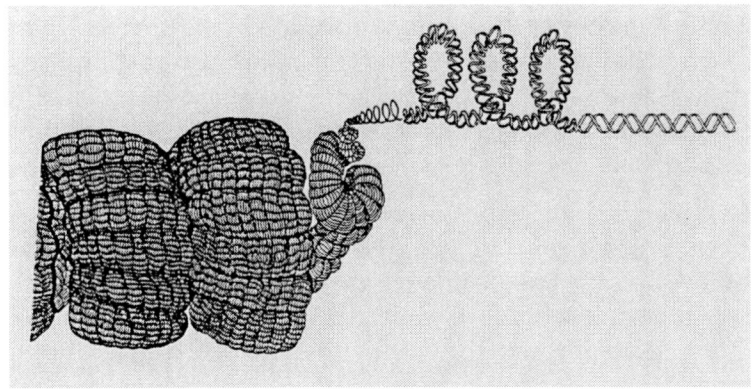

Life-body to enable the consciousness of human life to be apparent in the physical realm of reality.

It may seem complicated having two identical bodies. If the physical is part of the extra-physical why are we here in a physical body anyway? The Life-body seems more alive, lower maintenance and virtually indestructible. Why are we here in a body that is fragile and subject to disease, suffering and ultimate annihilation; a body that puts a dull on life. The answer may lie in all of these things.

Assuming the Life body grows alongside the physical, from conception to grave, with its genetic propensity and environmental influences, the physical body could enable the consciousness in the Life-body an opportunity to benefit from valuable lessons. The physical situation provides learning through difficulty, challenge and adversity. I believe the physical is both a womb and a school for new individualisations of consciousness, each with the potential for continuous living in the Living Universe.

Maybe the physical body is just a biological machine with a genetic operating system, capable of locomotion and self-replication, with sensors and a brain like a computer. I like to think of the physical world is a flight simulator for pilots of consciousness.

# Chapter 6 – Genesis of the Life Body

I believe the key to the human condition in the physical body lies in the limitations that the physical brain imposes on consciousness. In *The Doors of Perception*[3], Aldous Huxley summed this up by describing the brain as a filter of consciousness.

I suggest the limit in our awareness, that Aldous Huxley speaks about, is a critical part of a decision making process we are all involved in. I believe while we are alive in the physical world we have to decide where we want to go and who we want to be with after we pass through the gate of physical death on in continuous living.

I have believed this to be the case for many years but really came to appreciate it from the vivid descriptions in Howard Storm's book[4]. That is when I decided to write this book. I wanted to defy the denial of the soul promulgated by Democritus and other philosophers, by presenting a scientific model for the soul based on NDE descriptions and the Vortex theory. I also want to warn of the imperative of choice. The way we choose to live now could determine our long term destination in the future.

Most people don't like to think or talk about death. They either fear it or feel uncomfortable about it. Few people welcome it but it is inevitable for us all. Apart from birth it is the singular most important and unavoidable thing that ever happens to us. If we are too busy in our lives to consider the subject of death and the implications of near or actual death experiences we are fools because whereas NDEs usually paint a picture of the transition we call death somewhat like a happy homecoming after a day at school. Howard Storm's NDE painted a very different picture. His homecoming was terrible. He was met with a reception from hell.

3   **Huxley** A, *The Doors to Perception,* Penguin Books, 1959
4   **Storm** H, *My Descent into Death,* Clairview 2000

# Chapter 7

# A Reception from Hell

To begin with the voices were quite pleasant. They were calling from the hall, "Howard, Howard'. When he asked who they were and what they wanted they just replied, "Come, come quickly."

Howard said he responded by saying that he was sick and in need of urgent medical attention. They assured him they were there to help him, they could fix him but only if he came out to them in the hall. Howard was uneasy. It looked strange out in the hall. He had a feeling if he left the room there would be no way back but his wife and the other patient weren't responding and they couldn't help him anyway so he stepped out of the room and into the mist.

He saw shadowy figures a few feet away as in fog. They were male and female, pale and grey. When he moved toward them they backed away. Persistent in their urgency for him to follow them, the voices were no longer pleasant. In the semi darkness he was bewildered. There was nothing he could do but follow the shadowy figures.

As he shuffled along in his bare feet they crowded around him. He was communicating with them, not out loud but in his head. As questions arose in his mind about who they were, where they had come from and where they were going, they reacted with irritation and spoke to him in his head, insisting he hurry along.

His concerns and questions were meaningless, they said. The more he questioned the more annoyed they became. They assured him all his problems would be over once they reached their destination.

Howard recalled anxiety and perspiration on his journey but no tiredness and he felt very alive. As he hurried along with the ghostly crowd it grew steadily darker and the fog seemed to blacken. Then he turned to look back. He said he could see the

## Chapter 7 – A Reception from Hell

light in the open door and the room beyond with the body on the bed and Beverly sitting beside it in misery. It was surreal. Whenever he looked back the scene was still there but smaller than before and more distant.

Howard Storm said that in the fog there were no walls. The hall seemed endless and he felt the cold. There was no warmth normal to a hospital. There was no sense of the passage of time. He became more anxious as his companions rebuffed his pleas with sarcasm. They had become increasingly unfriendly and menacing as he progressed in the shadowy darkness. He wasn't dreaming. It was all terribly real and he began to realise he had been tricked into leaving the room and following these strangers into the dark.

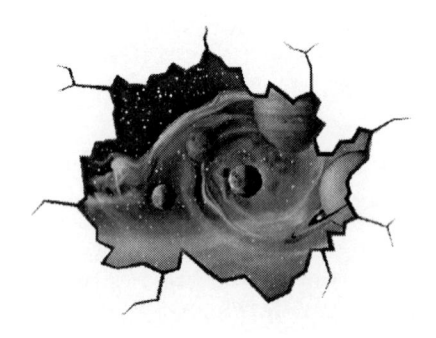

# Chapter 8

# The Shadow Lands

It is easy to pick holes in Howard Storm's account. He had his near death experience on the 1st of June 1985 and *Descent into Death* wasn't published until 2000. Fifteen years is ample time to forget things or spice up a story or even invent it. We only have his word for it anyway. The whole description reads like a horror story, too far fetched for fact. Many such arguments will surface in a critical and enquiring mind.

But what if Professor Storm didn't fabricate it? What if he wasn't a liar? What if he did have a real experience more or less as he described? Taking his serious academic background and his disbelief he seems the most unlikely person to spin a totally outlandish yarn. As a skeptic he would never have made a profound spiritual life change in a day unless something extraordinary happened to him and watching him on the Focus interview on the internet his sincerity is striking.

There is a saying that if the dead were to return from the grave, even then people would not believe. And that is precisely what is happening now. In NDEs and ADEs people like Storm are returning from the metaphoric grave to tell us of what happens when we die. We may say we are disbelievers but that is not true if we believe with certainty that death is the end of life.

What if death isn't oblivion? What if it is faith in death that is the delusion? What if people who believe in death wake up alive at the beginning of a fate worse than death?

Fools focus on how wrong something may be while the wise are more concerned at how wrong they may be!

In the shadow lands of despair Howard Storm was lost. He said that back in the hospital room he had longed for death to end the torment of life but he now realised death was no escape. He found himself tormented by a mob of unfeeling people forcing

## Chapter 8 – The Shadow Lands

him, mercilessly toward some unknown destination in the ever encroaching blackness.

He spoke of a deep sense of dread growing within him. The experience was all too real. He was more aware and sensitive than he had ever been in his life before. He was terribly alive and impossible things were happening that shouldn't be happening.

It was not a dream or hallucination. He wished it was. He was miserable and frightened, cold and lost. It was clear the help he had been promised was a ruse to trick him into following. He was reluctant to go any further, but any hesitation on his part brought on a hail of abuse and insults. He was told they were almost there, to shut up and take a few more steps. The terrifying darkness was complete when he decided to stop. That was when they attacked.

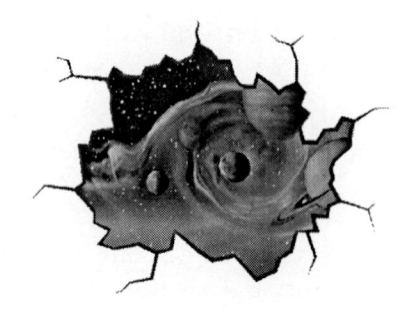

# Chapter 9

# The Attack

Professor Howard Storm was overwhelmed by the hopelessness of his situation when he stopped and refused to go any further. They came up to him shouting. He said he could feel their breath. They started shoving him and he hit back. In a frenzy of screaming they tore into him.

He fought back but that only exacerbated the attack. As he kicked and lunged at them they came back biting and ripping with nails that felt like claws. The pain was excruciating. Clearly, they were enjoying his suffering, laughing as they tormented him as cats would a mouse and then, slowly and methodically, they began to eat him alive.

Howard Storm said while he couldn't see in the darkness, every sound and every unexpected physical sensation registered with horrifying intensity. He realised the creatures were once human beings. His description of them was of the worst imaginable person stripped of every impulse of compassion.

He recalled they didn't appear to be controlled or directed by anyone. They were just a mob of beings driven by unbridled cruelty. He remembered having intense physical contact with them in the darkness as they swarmed over him. Their bodies felt exactly as human bodies. During his initial experience with them they were fully clothed but in the physical intimate contact he never felt any clothing.

Some people reading this will be in denial that it could be true but the details Howard Storm gave in his vivid description of the attack help me to understand the nature of the Life-body. Relative to the creatures that attacked him he was in a tangible body and theirs were very real to him. He felt their bodies as they crawled over him and their teeth and nails as they tore into him but he didn't die in the attack even though he was being torn apart.

# Chapter 9 – The Attack

Taking these descriptions at face value I would suggest the Life-body was very real in its own level of energy. It is impossible to say if it was atomic matter or more like plasma in a matrix; an alignment of non-atomic electric charged particles in an electromagnetic field template. If the template persisted and there was no lack of particles, theoretically the body could reform like iron filings in a magnetic field.

A physical body would not have survived such an attack, the equivalent of a pack of wild dogs, whereas in his Life-body Howard Storm survived. I realised the Life-body would be more proximate to the electromagnetic Life-field detailed by Professor Harold Saxton Burr[1], so more easily sustained by its template.

To understand the Life-body we have to reconsider our understanding of Life. The problem we have, as always, is our limited appreciation of reality. We have an unfortunate habit of placing a lot of trust in what little we know rather than allowing for the possibility that we might not know everything. We keep forgetting there may be things we have yet to discover that will change everything we think we know about the Universe.

Because we are biological life forms on Earth based on carbon atoms and molecules of water we limit Life to atomic and molecular biology. We believe Life is subject to death because life systems based on atoms don't last forever but physics is altering our perception of Life and persistent non-atomic Life could exist in space.

1   **Burr** Harold Saxton, *Blueprint for Immortality*, Neville Spearman, 1972

# Chapter 10

# The Living Universe

In the vortex theory I suggest that Life is based on electricity. Upon that premise I have come to the conclusion that the Universe could be alive. Let me explain myself.

Non-atomic matter, in the form of free moving charged particles, is electrically active. By contrast most atomic matter is electrically neutral. Atoms conduct electricity when they are ionised; that is when they gain or lose electrons. If Life is based on electricity then it is obvious that non-atomic matter would be a better environment for Life than atomic matter.

99.9% of the Universe is non atomic matter. For the most part this would be hot plasma in stars and cold plasma in space. Cold plasma consists mainly of electric particles blown from stars in solar flares and cosmic radiation. If the bulk of the Universe is electrically active and electricity is the basis of life, then the Universe could be a living entity. With the electric force of Life transmitted by plasma, space could be a full of non-atomic, non-organic forms of Life.

Our only experience of Life is in organic, carbon based biological organisms. Biologists define Life in terms of biology; the properties of biological things like locomotion, reproduction, respiration etc. Physicists would define Life in terms of physics; the properties of physical things like quantum activity, electricity and electromagnetic fields.

A biologist approaching Life would limit it to the Earth. A physicist approaching Life would see the potential for non-biological Life throughout the Universe. Physics is the king of sciences, not biology. When it comes to Life my advice would be to put the opinion of a physicist like David Bohm before that of a biologist like Richard Dawkins.

# Chapter 10 – The Living Universe

In his book[1], Mark Heley pointed out that Bohm, considered the possibility of non-atomic, plasmic forms of life. He frequently remarked that he had the impression that plasmas were alive and that they had many of the properties of organic life. The possibility of plasmic life is speculative but to quote Mark Heley from his book,

*"A much wider range of environments found in our universe suit the conditions for plasma-based life forms to exist in, than those that support carbon-based life forms; 99.9 percent of matter in our universe is plasma. Inorganic life could exist in much hotter and cooler conditions than organic life, such as solar atmospheres, planetary ionospheres or interstellar space."*

Mark Heley went on to cite a leading Russian scientist, Professor V. N Tsytovich, member of the Russian Academy of Sciences, who has done a lot of research into the self organising ability of plasma when exposed to electric charge. Tsytovich has developed his observations into a theory of inorganic life. He claims that his research has pointed to prime locations of plasmic, inorganic life around stars and in interstellar space.

Tsytovich has discovered that, in a gravity-free environment, plasma forms string like filaments that twist into helical strands resembling DNA. The strands are electrically charged. They are attracted to each other and assimilate less organised plasma. They have the ability to grow from this 'feeding' process and to reproduce by filament splitting. According to Tsytovich, *"…they are autonomous, they reproduce and they evolve."*

He claims that plasma can exhibit the criteria of what constitutes life. As Mark Heley says in his book, *"The conclusion of Tsytovich's report is groundbreaking in scope. He not only believes we need to widen our definition of what comprises life to include plasmas or inorganic life, but also that organic carbon-based life itself may possibly be derived from those plasma life forms. Tsytovich and other scientists*

---

1   **Heley** Mark, *The Everything Guide to 2012,* Adams Media. 2009

*have proposed that plasma life forms may, in fact, have been responsible for catalyzing the development of organic carbon-based life on Earth."*[1]

In *Dark Plasma Theory*, Jay Alfred supports this view. He cites Mircea Sanduloviciu of Cuza University in Romania as providing evidence of the connection between self organising plasmas and biological life on Earth. Many scientists, especially in Russia and Eastern Europe, now believe Life to be far more universal than carbon based organic life allows. We lag behind them in the West.

Despite the exponential advance of science and technology since the World War II we still fall in the trap of thinking we are at the end rather than the cutting edge of advancement. At the end of the nineteenth century, when horses were still a major mode of transport scientists thought they were at the peak of technological advancement and were approaching the end of discovery.

Science is never complete as a body of knowledge. It is always poised for new discoveries and prepared for the sight of a fresh frontier or another shock of the unknown.

It may be beyond our comprehension to imagine space full of Life on many dimensions and that the Universe may be a non-biological living being but if Life is defined in terms of electrical activity and the Universe is predominantly electrically active, non-atomic matter then all this is possible.

People will say I am speculating but speculation is the rocket fuel of science. Expanding our perception of Life beyond biology may be the only way we can make sense of near death experiences.

I believe Life on Earth is an end point rather than a start point of Life. This makes sense if we consider the dependence that Life on Earth has on water.

If life is defined in terms of electricity, atomic matter would be bottom of the pile of useful stuff for Life. For a living Universe full of plasmic life, atomic matter would be as dust, hardly worth the notice. For electric Life, electrically inactive atomic matter would present the ultimate challenge, the final frontier to colonize.

## Chapter 10 – The Living Universe

If life is based on electricity then Life would require atoms capable of conducting electricity. Ions do that. A medium for electric conductivity is water. Water is full of ionised atoms. Water, due to *hydrogen bonding*, dissolves salts into ions, which increases its conductivity.

Salt water is the best form of water to conduct electricity so it is the best medium for Life. Earth is a watery planet and life emerged in the sea. Life emerging in salt water supports the premise that electric activity rather than carbon is the key to Life.

Every living organism on Earth, from bacteria to humankind, is formed out of drops of sea water, suspended in protein jelly. We call these membrane wrapped watery droplets, cells.

Membranes are also in the cells. They are vital to the electrical activity of biological Life. Electricity depends on potential gradients which are possible through varying concentrations of ionised chemicals across membranes. The electric theory for Life makes sense of the dependence of biological life on water. However, future historians may compare biologists who restrict Life to water dependant Earth bound organisms to the 19th century naturalists, in the age before electricity, who viewed carrier pigeons as an advanced form of communication.

Biology depends on electricity. Electricity does not depend on biology. There could be systems supporting the electric activity of Life as advanced on biology as the internet is an advance on the carrier pigeon as a means of communication.

From our position on Earth, carrier pigeon in hand, it is impossible to even imagine the systems of non biological Life that could have evolved in the countless galaxies, each with hundreds of billions of stars in the billions of years the Universe has been in its current phase of existence.

My limited view is that in space plasma might have somehow become organised into electromagnetic fields capable of holding potential gradients. These could then have streamed subatomic particles into formations of matter. Atoms could represent just one of these forms. There could be many others. These electromagnetic plasma matrices might be able to form bodies similar to our own. Our bodies are, after all, only combinations of subatomic particles.

# Continuous Living

We are all familiar with the electrical activity forming two dimensional realistic images on television screens. Take that a step further and we could develop a technology for three dimensional realistic images. The next step would be to project our consciousness into them so we can step into the drama and be immersed in the virtual reality. That might be possible if we got to grips with plasma technology, but let's turn the logic around.

If Life has been evolving in space plasma for billions of years in trillions of star systems, in countless galaxies maybe consciousness has already projected into worlds along the lines we can only imagine as 3D virtual realities.

Maybe we are one such projection? Maybe there are other worlds existing, not with biological life, but with Life manifest through forms based on a plasmic, electromagnetic field matrix. Maybe living bodies of electric particles exist in these worlds as non-atomic, electromagnetic field matrices, which may also exist in interstellar environments.

Maybe biological Life on Earth is a place in the Living Universe where plasmic bodies in extra-physical matter are formed because it could be that the physical body provides scaffolding for their genesis and biological life provides an ideal environment for their development.

Howard Storm thought he had just one body. He had no idea as a kid that he was growing in two bodies not just one. Then when he nearly died and the body he didn't know about separated from the one he did, he found himself in a situation where the decisions he had taken in his physical life influenced his onward destiny. Having rejected all possibility of continuous living and embraced the darkness of death it seems he rejected himself into darkness and ended up with the living dead.

When Howard Storm was attacked, according to my premise, he was in an electromagnetic *feeling body,* the same body his consciousness felt through when he was alive in his physical body. He was still alive but without his biological body he was lost to his wife and family, his properties and his career. The only thing he was left with was his pride.

## Chapter 10 – The Living Universe

Thanks to his position in academia and the attitudes that went with it, his knowledge and his confidence in consensual scientific opinion, through his pride, he had chosen the polarity of death and denial that left him exposed. After the transition we call death, he was exposed to a hostile situation totally foreign to him. He was in darkness, at the behest of a terrifying and merciless mob. To him they were very real as they tore into him. They would not have been real to you and me in our atomic bodies but if Howard Storm was in a plasmic body like them, he would have felt them keenly as he did because of relativity!

# Chapter 11

# The Hermetic Teaching

I have attempted to explain Howard Storm's experience on the basis of three fundamental levels of energy in the Universe, the physical, the extra-physical and the super-physical each delineated by a higher relative speed of energy, which is a subset of the one above, as depicted in the model of concentric spheres. This arrangement of the Universe in levels or planes was first described by a teacher known as Hermes in Greek and Mercury in Latin; names that meant movement in their respective languages.

According to Augustine, Hermes Trismegistus was a real person who lived in Egypt three generations after Moses[1]. The Greeks turned Hermes into a god much as the Romans did with Christ. Religious cults were created around their persons. Democritus was challenging the religion of his day by rejecting the teachings of Hermes as do people today reject the teachings of Christ in reaction to religion.

The damage done by the Greek and Roman religions in deifying these men is unfortunate as both of them dedicated their lives selflessly to the upliftment of humanity and their teachings are eternal. We dismiss them at our peril!

Hermes is considered the father of science and medicine. His symbol, the *Caduceus* is still in use today in medicine. Hermetic alchemy led to chemistry. Lord Rutherford, the father of nuclear physics, honoured the tradition in science of respect for Hermes by including him in his heraldry. Sir Isaac Newton revered Hermes.

Hermes taught that the Universe is a mind. He also introduced the principle of hologram symmetry. What we describe as

1   **Augustine**, *Civitas Dei* XVIII.29

## Chapter 11 – The Hermetic Teaching

fractals, he worded as: *As it is above so it is below: As it is below so it is above.*

Hermes' most significant statement, from which his name is derived, was that *everything is movement*. In that teaching Hermes anticipated Einstein and the understanding in Quantum theory that everything is energy.

Hermes also taught that polarity exists at every level in the Universe. He emphasised that the rhythm of coming and going, growth and decay, life and death, expansion and contraction, is universal.

The Hermetic teaching that every cause has an effect is embodied in Newton's laws of motion.

Hermes also contended that the principles of masculine and feminine are found at every level in the Universe. I have used this principle in *The Vortex Theory* to account for wave-particle duality, kinetics and quantum states in the atom as well as the flame and ionic and covalent bonding in chemistry.

The teachings of Hermes are seminal in my work. They enable me to explain near death experiences from the symmetry of *As above so below.* The principle, that patterns are the same in microcosm and macrocosm, suggests the Universe is structured much like the modern atom. The planes described by Hermes could represent a progression of higher critical speeds of energy in the *macrocosm* corresponding to the structure of the atom in levels or quantum states, based on a progression of higher frequencies in the *microcosm*.

For Hermes spirituality had nothing to do with religion. It was part and parcel of his science. So it is for me, I am not attempting to resurrect religion; I am endeavoring to restore to science teachings of Hermes that were dismissed by Democritus and misinterpreted by Aristotle, Ptolemy and other philosophers in ancient Greece and Rome.

The levels of reality Hermes spoke of, which relate to soul and spirit, fit my descriptions of the extra-physical and super-physical levels of quantum reality. The words soul and spirit are relingos that evoke a strong reaction of association with religion in most people. I am endeavouring to address the principles the words represent from a standpoint in science.

An objection may be raised that the idea of a soul that can exist apart from the body is no longer relevant in the modern era. But that objection comes from the materialistic philosophy of Democritus. NDEs suggest that the ideas of soul and spirit surviving physical death were relevant in the past, are relevant today and will always be relevant.

Hermes also taught that reason wills everything into existence. As creatures with free will, it would seem we have the ability to reason our own future destiny and influence the future destiny of others.

Hermes likened us to the divine principle described in ancient Egyptian as *Hu*. That is why he called us *Human* beings. Hermes taught that with creative and destructive minds, we in the *As Below* reason things in and out of existence in this world much as the Universal mind, in the *As above*, reasons things in and out of existence in the Universe.

I believe each of us determines our fate in one or other polarity of the Universe according to our exercise of reason; as Storm discovered in his NDE. He ended up with other human beings who had exercised their reason during their lives on Earth in much the same manner as he had and like them he had to endure the reality of another relingo *sin*.

Most of us think sin is doing wrong but sin literally means to be cast out or rejected. Being bad may not be the prime cause of rejection, or sin, as most people assume. Rejecting Life and choosing to believe in death may be a cause because our rejection of Life could lead to a reciprocal rejection. Life, respects our choice. This is not religion, it is physics; the law of cause and effect in action, as Ian McCormack discovered to his cost.

# Chapter 12

# The Wipeout

In the language of surfers, you have a *wipeout* if you fall off a wave and get mashed in the break. In 1982, when he was only 24, a surfer from New Zealand called Ian McCormack had the ultimate wipeout; but it wasn't falling off a wave!

Ian McCormack wasn't wicked. Far from it, he was a nice guy. Surfers usually are nice guys. They are handsome, healthy, wholesome young men who don't rob or murder, or gamble or drink excessively. I don't think I have ever met an evil surfer. I come from a family of surfers. The image that Ian was so evil he deserved hell doesn't fit. There is no devilish thing that Ian McCormack did to deserve his fate; so what happened?

Ian was swimming off a reef in Mauritius at night, hunting crayfish with a torch. He had on a wetsuit but unfortunately his forearms were exposed. I remember at his age I used to surf in Cornwall with skimpy wetsuits because I couldn't afford anything half decent; but then I wasn't in danger of encountering Box jellyfish.

Practically every year people swimming in the tropical waters of the South Pacific die of Box jelly fish stings. Often after just one sting they die within minutes because the Box jellyfish venom is regarded as the second deadliest in the world. In a night of foreboding storm, Ian was hit across the forearm by the destroying tentacles of five of the most dangerous jellyfish in the world.

He managed to swim back to the boat and one of the local lads, who had gone with him, poled him back to the shore. He helped Ian to a roadside but then abandoned him to his fate in order to go back and collect his brothers from the reef.

With the venom seeping into his bloodstream, Ian was dying. He managed to limp to a restaurant where he saw taxis. He begged the drivers to take him to a hospital. They asked if he had money.

# Continuous Living

Obviously he didn't so two of the drivers walked away. The third pulled him into a taxi but then pushed him out onto the street outside a tourist hotel because he couldn't pay the fare to the hospital.

Fortunately a black security guard recognised him and carried him into the hotel, dropped him in a cane chair opposite the Chinese owner of the hotel who was gambling with a couple of companions, and ran off to call an ambulance. Ian begged the Chinese to drive him to hospital but they rebuked him for taking heroin and continued their game.

He wanted to get up and hit them but the stung arm, swollen and covered in burning weals, had gone into spasm and muscles in his body were reacting with jerks in response to the creeping poison, throwing him about in the chair. Then he couldn't move as a deadly cold crept through his limbs.

Eventually an ambulance arrived. At first it drove off because he was white and it was for blacks, but then his black friend managed to whistle it back. Eventually, loaded on board the rickety Renault apology of an ambulance he was driven off to a dilapidated hospital. Unfortunately it was already too late.

In the hospital a doctor recognised his symptoms immediately and ordered a nurse to set up a drip. Another took his blood pressure but it was too low to register. A third nurse attempted to inject antitoxin but the vein just blew up like a balloon. The antitoxin didn't go anywhere because the circulation had virtually stopped and the blood wasn't moving sufficiently to get it into his system. Ian was still conscious. His eyes were open and he could see the activity but he was immobile. Then all of a sudden he let out a sigh and passed away. Ian said that while he was being treated he kept feeling like floating away. It was when his body let out the sigh he realised he had floated into darkness.

The hospital checked for signs of life but his heart had stopped and he was no longer breathing. The doctor pronounced him as dead and he was transferred to the mortuary.

Skeptics may argue that Ian was paralysed by the jellyfish venom and in a state of suspended animation he was hallucinating due to the poison. The fact that he came round in

## Chapter 12 – The Wipeout

the mortuary some twenty minutes later could be used to support that view.

The fact is Ian McCormack was recorded as clinically dead. There were no signs of life. He had an actual death experience. He had the same two body dilemma as Howard Storm and like Howard he went somewhere that wasn't very nice.

He didn't know he had died because he felt very much alive and his senses were heightened but he was in total darkness. He was standing in a huge broad place. It was like a vast hall in a blackout. He had the sensation of being in his body but couldn't see it because the darkness was so complete.

He reached out to see if he could find a wall switch to turn on the lights but there was nothing there. It was bitterly cold and he was aware of a brooding presence. He knew it was aware of him. It felt invasive and evil. Then he became aware of other people milling around in the same predicament of fear and confusion he was experiencing.

When he started wondering where he was one of them shouted "Shut up!" in his head. Ian was experiencing the same mind to mind communication as Howard Storm.

Ian realised the people were aware of his thoughts and were responding to them in his head. Someone else shouted, "You deserve to be here!" A third screamed at him that they were in hell. Ian said he had always imagined hell as a fun place where everyone could do anything they liked but now he was there he found it was terrible and frightening. No one could do anything. There was no end to the space, no sense of time, and there was no way out.

# Chapter 13

# The Holographic Universe

When reporting the near death experiences of Howard Storm and Ian McCormack I took the trouble to suggest ways of explaining away their experiences. I did this to provide a way of explaining away the premise I am presenting, in case that is what someone wants to do. The most important thing is freedom of choice. None of us wants to be told what to do or what to believe and the last thing we want is a threat of hell hanging over our heads, but for those worried in case hell is for real I offer the suggestion of risk assessment.

Can we afford to take a risk with hell! What if hell is true? What if those two men really did have an after death experience of hell? No one can be absolutely sure either way!

My attitude is it is better to be safe than sorry. It is a bit like fitting smoke alarms. We could say "Oh it will never happen to me!" and not bother with fitting the alarms or we could say "I hope it won't happen but in case it does I will take action!" In this case it would be to fit smoke alarms.

If you would prefer to play safe rather than play Russian roulette with hell perhaps it would be prudent to treat the warnings of these death experiences as true unless proven false.

No scientist can validate the experiences of Storm and McCormack. I have taken them on face value because they point to a powerful idea in my line of scientific enquiry. I think Howard Storm and Ian McCormack had an experience of the immune system of the Living Universe that kicks in after death. That is only a premise. If you ask me to prove it obviously I can't.

My idea that the Universe is alive is only a premise. But if it is a body with an immune system we could understand it in terms of a hologram. Our body could be a fractal in a holographic Universe. It doesn't matter how big or how little the fractal may be in the set, it is of equal significance because the whole is made

## Chapter 13 – The Holographic Universe

up the parts; they are all connected. The hologram is nothing but a collective of its fractals. If *Chaos theory* is correct and the Universe is holographic and we are holograms of the Universe then each of us would be images of the Universe and the Universe would be an image of us. Just a thought; the recently discovered walls of galaxies could be membranes!

I don't believe there is a God presiding over us, judging as to whether we are good or bad and deciding whether to send us to hell or heaven. I believe in Life and I believe that Life is based on electricity. I don't believe in Judgment; I believe in electric switching. I reckon that if the Universe is electrically live it will run like an electric circuit with switches that are sensitive to polarities and barriers that are sensitive to frequency. I reckon the Universe is like the computer I am writing on, and we are as electric particles running around in it, directed in our progress according to our polarity and our frequency. My beliefs are based on physics, not religion.

I treat the word, *God* as a relingo. It is charged with centuries of connotations but it is the name used by a lot of people to denote the Life principle of the Universe. We all have names. Mostly they were laid on us by our parents. We may not be happy with the name imposed on us but if everyone is determined to use it as our call up signal, we have little choice but to respond to it.

Whenever anyone shouts "David" I always react, even if it isn't me they are calling. Even though I didn't choose the name I respond to it. If I hear someone proclaim they don't like David, they don't believe in David and anyone else who does so is no better than someone who believes in the tooth fairy I am hardly likely to invite them to my party.

The idea that the Universe is alive and sensitive to our thoughts is only speculation. It is only a premise that there are frequency switches, operated by the polarity of our mindset when we pass through the gate of death, that direct us like binary in one of two directions. But I am trying to make sense of the death experiences that Howard Storm and Ian McCormack went through in terms of my understanding of physics, in case they are true.

# Continuous Living

My concern is that neither of them was particularly bad. They actually seemed quite normal and it may have been that being normal was their problem. It is normal these days to reject God and religion or to be indifferent to them and to treat selfishness as acceptable behaviour. You see if God or goodness were just a tooth fairy variant it wouldn't be a problem but if God or the principle of goodness is the name we have given by consensus to the Life principle of a thought sensitive, holographic Living Universe we happen to inhabit then rejection or indifference to this principle might be a bit risky!

The Howard Storm, Ian McCormack NDEs suggest if we close our minds to the possibility of continuous living and believe in total darkness at death then total darkness could be what happens to us as it was the option of our choice. By rejecting the possibility of a reality beyond the physical we may exclude ourselves from it. By rejecting Life we may be rejected by Life. No one wants to be rejected. We all deserve light and love and unending joy in continuous living but it could be that in order to attain it we may have to believe in it and desire it.

Considering the way Howard Storm was cast out into darkness I am nervous in case his atheism was linked to his rejection. I find it hard to believe the Universe would be concerned with our beliefs apart from the extent to which they cause us to hate each other. The lack of love between believers, different believers and non-believers leads to hell on Earth, as we witness in the middle East, with continuity in continuous living. To that extent I am concerned at the growing popularity of atheism, as a potential source of more conflict with people, even more so at its being promoted as scientific in schools and universities.

Howard Storm's professor of philosophy converted him to atheism when he was a student. It is not that I want to promote belief in God, as I resonate with many of the anti-religious sentiments of atheists and I enjoyed *The God Delusion* by Richard Dawkins[1]. Nonetheless Dawkins heads a movement promoting

1　**Dawkins** Richard, *The God Delusion*, Bantam, 2006

## Chapter 13 – The Holographic Universe

atheism encouraging people to believe in nothing but the oblivion at death and disallowing the possibility of continuous living.

Richard Dawkins is a biologist and the termination of biological life is obvious to biologists. Because materialist scientists like Dawkins restrict Life to biology, they have no reason to believe in the possibility of continuous living. Biologists don't entertain the possibility of inorganic Life as inorganic Life is in the domain of physicists not biologists. Also I believe many scientists deny the evidence of NDEs for continuous living because NDEs contradict the materialistic scientism that supports atheism. The problem as I see it is the tendency of atheists to deny the existence of Spirit and the Soul and to lump continuous living in with religion.

Atheists like Richard Dawkins detest religion and despise the idea of God, and with good reason. The God of the Bible is evil! Jesus Christ knew that. He said the Old Testament God is the father of lies and a murderer from the beginning. Jesus[2] warned there is no truth in him and yet Western religions, even those in Christ's name, have set up the murderous father of lies to be worshipped and glorified.

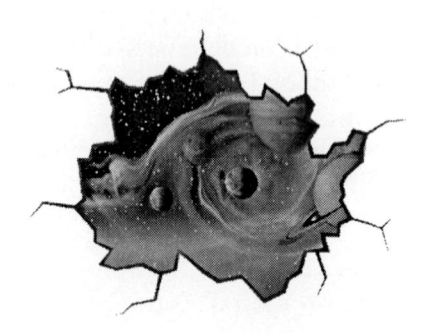

2   John 8:44

# Chapter 14

# The Evil God

In The God Delusion Richard Dawkins wrote: *"The God of the Old Testament is arguably the most unpleasant character in all fiction: jealous and proud of it; a petty, unjust, unforgiving control freak; a vindictive, bloodthirsty ethnic cleanser; a misogynistic, homophobic, racist, infanticidal, genocidal, filicidal, pestilential megalomaniacal, sadomasochistic, capriciously malevolent bully."*[1]

In my book *The Role of Evil in Human Evolution*[2] I detail why I believe the God of the Bible is corrupt and evil. The Bible tells of God appearing to a desert nomad called Abram. God made the tribesman an offer that his progeny would inherit the earth in return for worship and unswerving loyalty. Abram was very impressed.

Succumbing to the temptation, he bowed down and worshiped God. In return God renamed him Abraham and asked for circumcision of all his male descendants. Thousands of years later we see the impact on the Earth and world affairs of the descendants of Abraham, the Israelites and the Arabs!

The Bible goes on to tell that Jesus Christ was tempted in a similar manner. An apparition appeared to him offering him the kingdoms of the Earth in return for his worship. Jesus refused the offer and rebuked the apparition with the words, *"Get behind me Satan."* (Matthew 4:10) Whether that apparition was the same one who came to Abraham, is not certain but the temptation was certainly the same and religions encouraging worship of the God

---

1  **Dawkins** Richard, *The God Delusion*, Bantam, 2006
2  **Ash** David, *The Role of Evil in Human Evolution*, Kima Global Publishing, 2007

## Chapter 14 – The Evil God

of Abraham include Judaism, Islam, Catholicism, Christianity, Mormons, Jehovah Witnesses and Freemasonry.

The leaders of Judaism had Jesus killed but the Romans deified the Christ they crucified on behalf of the Jews. In AD 325, at Nicea – in what is now Turkey – the Romans established the Roman Catholic Church. Anyone who dissented from the new religion was branded heretic, hounded down and killed.

When a group of pacific vegetarians set up their own sect in the Languedoc region of France, they were slaughtered mercilessly for their heresy. These were the Cathars. The Cathars taught that the Old Testament God is Satan. They attributed this article of their faith to the Apostle John. He got it from Jesus and like Jesus they were murdered.

In verse 44 of chapter 8 in John's gospel Jesus is reported as saying to the Jews:

*"You are from your father the Devil, and you choose to do your father's desires. He was a murder from the beginning and does not stand in the truth, because there is no truth in him. When he lies he speaks according to his own nature, for he is a liar and the father of lies."*[3]

The word *Devil* it is derived from the old French for *'of evil'* and *evil* is *live* spelt backwards. Evil is anti-Life. Evil is belief in death.

Jesus Christ didn't speak of God because that word didn't exist in his day. He spoke of the *father*. He made it clear that the Jews were following the father of lies and yet the Roman Catholic Church established a trinity in which the Old Testament God, declared by Jesus to be the Devil, is placed in the position of father. Jesus is set below him is a subservient son and somewhere flapping between them is a Holy Ghost!

So where did the relingo *God* come from? The word God comes from the Old English word for *good*. It is still the word for good in Scandinavian languages today. Its use in the religious context

---

3  **Anglicized Bible** *New Revised Standard Version* Collins 2011

with 'g' in the upper case became commonplace thanks to the Tyndal and King James Bibles.

A problem with the relingo, God, is that it is singular. If you read the book of Genesis you will come across a number of verses where God is presented in the plural. Where a verse begins: *And God said "Let us...."*

God appears to be speaking as a group rather than as an individual. The reason for this linguistic confusion is that in the original Hebrew texts, upon which the Old Testament was based, the scribes didn't use the word God. Where modern Bibles have God or the Lord they had the word *elohim*.

In Hebrew the word *el* is the equivalent of the English relingos god or divine. *Elohim* is the plural of *el*. If it were an accurate translation of the original texts, the English Bible would read: *And the gods said "Let us...* The problem arises not from the translations from Aramaic or Hebrew into Greek then into Latin and on into English, but from the transition of the Hebrew religion from polytheism to monotheism.

Before Abraham there were a number of gods but monotheism came into being when Abraham swore allegiance to the one we now call God. I believe Abraham's monotheism was a coup by one of the elohim to push the other members of his team out of the picture in order to seize control of the planet and dominion of humanity.

The Bible tells of Abraham's loyalty being tested by God. He was asked to offer his son as a human sacrifice. At the point when he was about to kill his son, an angel appears and calls on him to stop. If you read the story carefully you will notice the angel speaks to him as God. There is confusion in the Bible between God and the angels! *El* at the end of *angel* denotes divinity in the Hebrew. The relingo, *angel*, means *divine messenger*. Angels are visitors; they don't belong to the Earth so by definition they are extraterrestrial.

If we allow for the possibility of non-organic Life filling space then there could be visitors from outer space. Plasmic forms of Life could abound and they could have been visiting this planet as *plasmic holographic inserts* for many thousands of years.

# Chapter 14 – The Evil God

If they do exist and if they do visit the Earth we don't know who they are, where they are from or what their intentions are. The Elohim could have been a group of extraterrestrials. Whatever relingo we use, be it *god*, *divine* or *angel* is irrelevant. If extraterrestrials do exist it would make sense to treat them with caution. Stories like the temptation of Abraham by an extraterrestrial apparition should set our alarm bells ringing. Certainly we shouldn't have blind faith in them!

Richard Dawkins presents Abraham's God as fiction. In my opinion, that premise is dangerous because it lulls us into a false sense of security. Monotheists who follow Abraham tell us to believe in God. Atheists who follow the likes of Richard Dawkins tell us to disbelieve in God. I suggest that as the Universe may be full of non-organic Life, forms of Life we haven't even begun to comprehend, it might be foolhardy to dismiss reports from ancient or modern times of alien Life forms visiting us from space.

I am of the opinion we should treat them as real so we can treat them with caution. I think Abraham was foolish to bow down and worship an alien and monotheism has a lot to answer for in making out that his foolish and reprehensible act was good.

# Chapter 15

# Monotheism

Monotheism emerged as an organised religion in ancient Egypt. Archeological discoveries reveal that 3350 years ago there was a glorious city called Akhet-Aten which was the capital of the Egyptian empire then and home to the Pharaoh Akhenaten. Rabbis Messod and Roger Sabbah wrote, in *Secrets of the Exodus*[1], of Akhet-Aten as:

*"...a city of palaces, temples and obelisks covered with gold as far as the eye could see...The art, the beauty and the refinement of Akhet-Aten reflected the very height of Egyptian civilization...people had received from Pharaoh the immense privilege of coming with their families to live in this sacred city, the Holy City of God Aten. By order of Pharaoh, they had converted to the new religion, the new cult of the One God, to monotheism."*

The idea, that Akhenaten was the pioneer of the monotheistic religion, which later became Judaism, was first suggested by Sigmund Freud in his book *Moses and Monotheism*. Freud argued that Moses had been a priest of Aten forced to leave Egypt with his followers after Akhenaten's death.

According to the Bible, monotheism originated with Abraham and when his great grandson Joseph gained rank and influence in the court of a Pharaoh, the God of Abraham was introduced to Egypt. There is no independent evidence to support this but it might explain how monotheism came to be embraced by a Pharaoh and why Moses was in the court.

However the records of ancient Egypt contradict the Bible story of the monotheist Hebrew people being held in slavery by a heartless Pharaoh and there is no mention of great plagues

1    SSabbah M & R, *Secrets of the Exodus*, Thorsons, 2002

# Chapter 15 – Monotheism

bringing ruin to Egypt. Messod and Roger Sabbah suggest it was the people of Egypt who were enslaved by Akhenaten to support the cult of monotheism he endorsed and Egypt was brought close to ruin by the amassing of wealth and resources to the city of Akhet-Aten.

Other Egyptologists confirm this view that Egypt went into decline due to the neglect of the monotheistic Pharaoh. After Akhenaten's death, and that of his successor two years later, the younger son, nine year old Tutankhamen, became Pharaoh. Tutankhamen's uncle Ay, acting as regent, evicted the monotheist cult of Aten in his concern to rebuild Egypt and then he restored the original polytheistic cult of Amen.

Roger and Messod Sabbah contend, from their extensive research, that the followers of Aten were settled in the Egyptian province of Canaan, with the sanction of Ay and the support of the Egyptian army, and were allowed to take the wealth of Akhet-Aten with them.

No evidence has surfaced to support the Bible stories of the drowning of the Egyptian army or the wandering in the wilderness. Akhenaten is a source of fascination and speculation and little is known about Ay but the rabbis, Roger & Messod Sabbah, present a convincing and well researched account of the Exodus suggesting that Judaism was a cult of monotheists from many nations that were turned out of Egypt rather than a separate nation of disenfranchised people. This suggests the Hebrews originated as a religion not a nation.

Exiled from Egypt but settled in the Egyptian province of Canaan, the monotheistic Egyptians passed the story of the exodus from father to son over many generations as an oral tradition. The stories were then written, centuries later, during the reign of King Solomon when the written Hebrew was established.

While the Bible story of the Exodus does not fit with the historical records of Egypt for that time it does record that Canaan was secured by genocide. The cult of monotheism was established by Akhenaten through temptation with women and wealth before the exodus occurred. All this is a reflection on the polarity of the monotheists and supports the accusation of Jesus

# Continuous Living

Christ that the father of the monotheist religious cult, founded by Abraham, was a murderer and a liar; that the one now called God was corrupt.

I believe the monotheism coup was by an entity with a vested interest in death because the followers of this monster built a place of blood sacrifice, a temple dedicated to death over the summit of Mount Mariah where Abraham offered his son for sacrifice.

Jesus Christ was never son of this revolting God; he led the revolt against him. He never called himself the son of God. He always called himself the son of man. I believe he could well have been a freedom fighter against Roman occupation and oppression in his youth. I can see no other reason why the Romans would have crucified him when he was handed over to them by the Jewish authorities.

To break with any association with those religions I have dropped the relingos Jesus and Christ and have reverted to the name the revolutionary was know by; give the man his own name is what I say; Yeshua ben Yoseph.

Yeshua was opposed to the religious establishment. He said the priesthood was a nest of snakes and he went berserk in the Temple turning over the tables of the money merchants. He didn't endorse monotheism because he spoke of two fathers, one good and one evil. Yeshua set out to rescue his people from the monotheistic religion that was burdening them with endless petty rules and regulations for right thought and behaviour. He wasn't rich and famous. He mixed with the beggars and prostitutes, the poor and the downtrodden.

His mother Mariam was a maiden when she conceived out of wedlock. She was possibly raped by a Roman soldier, which was common in the Roman provinces. That could account for why Yeshua might have been opposed to the Romans in his youth. The stigma of being illegitimate would have made it difficult for him when he became a healer and a teacher with his revolutionary message of unconditional love.

Yeshua spoke of continuous living, an idea also taught by the Egyptian Hermes and the Gnostics. He taught that those who lived in the light with love in their hearts would continue to live

## Chapter 15 – Monotheism

like that whereas those who chose darkness and death would live continually that way. His message was simple; Love Life and love who ever you are with.

After Yeshua was murdered at the behest of the monotheists his message of love and continuous living spread like wildfire throughout the Roman Empire. It gave hope to the oppressed and no matter how hard the Romans tried to stamp it out it grew from strength to strength.

Eventually, perceiving the threat of Love and Life to his dominion, Emperor Constantine deified Yeshua and Latinising his name to Jesu, harnessed the momentum of his movement to monotheism, morphing it into a new form of Roman domination; the Roman Catholic Church.

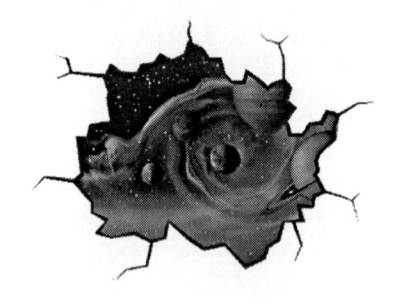

# Chapter 16

# The Dragon of Old

In *The God Hypothesis*,[1] Dr J. Lewels, spoke of the Mandaens, an early Jewish sect who believed in a polarised universe, divided equally into the worlds of light and darkness. *"To them, the physical world, including the Earth, was created and ruled over by the Lord of Darkness... variously called Snake, Dragon, Monster and Gian...thought to be creator of humanity."*

To quote Tobias Churton from *The Gnostics*[2],

*"There were Jewish schools, much given to speculation on the nature of God and the constituent beings which constituted his emanation or projection of being. Some of them appear to have been profoundly disappointed with the God of the Old Testament and wrote commentaries on the Jewish scriptures, asserting that the God described there was a lower being, who had tried to blind Man from seeing his true nature and destiny. We hear their echoes in some books of the Nag Hammadi Library, namely, 'The Apocryphon of John' and 'The Apocalypse of Adam. They believed in a figure, the 'Eternal Man' or 'Adam Kadmon' who was a glorious reflection of the true God and who had been duped into an involvement with the lower creation, with earthly matter, ruled by an inferior deity who, with his angels, made human bodies."*

According to the second century Gnostic mystic and poet, Valentius, the Old Testament God was a demiurge: *...determined his creatures (us) shall remain unaware of their source. (The Gnostics, p.55)*

1   **Lewels** J., *The God Hypothesis*, Wildflower Press, 1997
2   **Churton** T, *The Gnostics*, Weidenfeld & Nicolson, 1987

# Chapter 16 – The Dragon of Old

Cathar *Perfecti,* taught that *Satanas* made Man in his own likeness, trapping the souls of angelic beings into his human creations. As described by a Cathar in *Les Questions de Jean:* "*And he Satan imagined in order to make man for his service, and took the lime of the earth and made man in his resemblance. And he ordered the angel of the second heaven to enter the body of lime; and he took another part and made another body in the form of woman, and he ordered the angel of the first heaven to enter therein. The angels cried exceedingly on seeing themselves covered in distinct forms by this mortal envelopment.*"(*The Gnostics p.74*)

William Blake perceived Jehovah as having fallen from an original state of virtue and dignity as did Yeats:...*He is recognizably the 'Elohim' of the Old Testament in his 'aged ignorance', carrying the books of the Law, and yet Blake's Creator is not wholly evil; for as the 'Ancient of Days' on the frontispiece of 'Europe' (1793) we see the Creator with his golden compasses, who, though in part fallen, 'derived his birth from the Supreme God; this being fell, by degrees, from his native virtue and his primitive dignity'* (W. B. Yeats - *(The Gnostics, p.147)*

In a vision, Daniel (Daniel 7:9) saw the *Ancient of Days* seated on the 'throne' as the God of his fathers, Abraham, Isaac and Jacob. It is the description of Abraham's God as the Ancient of Days that gives me the clue to what God is. I reckon the evil God could be somewhat like a parasite!

Parasites are ancient forms of life that have faults which have been corrected by evolution in later types of organism. Let me give you an example. The earliest worms were tubes of muscle with a skin and a gut. These include flatworms such as tapeworms and flukes and round worms such as the nematodes.

The problem for these most ancient worms is that they couldn't eat without moving because the waves of peristalsis that enabled them to swallow their food also caused them to move. By the same token they couldn't move without throwing out their food because the waves of peristalsis that enabled them to move also caused them to evacuate.

Earth worms are a more modern model of worm in which there is a split in the muscle so that some of it lines the gut for swallowing and defecating and the rest of the muscle is under

## Continuous Living

the skin for locomotion. We are all based on this improvement. The muscle that operates our guts is separated from all the other muscles in the body by the pericardium.

The ancient worms that couldn't survive on the open range, managed to stay the execution of natural selection by living in the bodies of other creatures where they could eat without moving. They became parasites. Parasites destroy the health and can kill the host. Parasites are essentially evil because preying on healthy life they cause disease and death so, despite being alive, they are, in effect anti-life.

Other ancient organisms that prey on healthy life are bacteria and viruses and single cell organisms like the plasmodium that causes malaria and amoebae that cause dysentery. These are called pathogens. You may not see a pathogen and therefore not believe in it but it could still kill you. In fact not believing in it gives it cover while it destroys you.

If we consider the Hermetic symmetry of ...*as it is below so it is above* we would anticipate parasitic and pathogenic entities in the non-biological Life of space. If this is so, then we would also expect the pathogenic forms of non-organic, plasmic Life to be the most ancient. We humans, being the very latest models, could be target prey for plasmic parasites. As Ancient of Days, the God of the Bible, Mandaen Lord of Darkness and demiurge of Gnosticism could be an ancient form of plasmic life.

In Gnosticism and the Mandaen lore, Satan Lord of Darkness is believed to have made us as slaves. If that were so then he could have a claim on us which we would have to denounce by a conscious choice if we want to be free of him.

This is the basis of the religious concept that unless we make an effort to be saved we are the property of Satan and destined for hell. That is a terrifying thought. We could dismiss it if death were oblivion. But if we do live on after death then, as Ian McCormack and Howard Storm discovered, we might end up in hell by default with Satan in the dark unless we do something definite to avert that disaster.

Some people despise religions because they appear to be trying to make a claim on us. Religions contend they are trying to help us challenge someone sinister who already has a claim on us.

# Chapter 16 – The Dragon of Old

They say we have a free will which we can exercise to free ourselves but it is up to us to do it, either through our own devices or with the help of others. This, I believe, may be why the surfer Ian McCormack ended up in hell. Through indifference and selfish living he didn't choose heaven.

We may object to the way things are presented by religions but can we afford to be indifferent to the essential message of true religion that we be kind and considerate to each other. We may dismiss them in the name of science, but if we are in a Living Universe what they are saying may be just a reflection of the patterns of Life we observe in the world around us. Life is not sentimental but then neither are we.

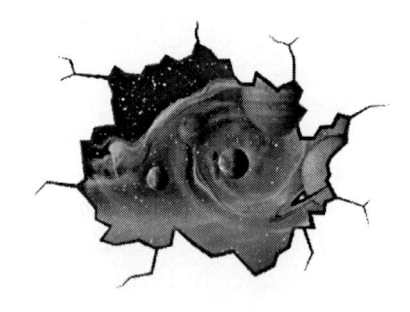

# Chapter 17

# Karma

In India the word *karma* is used for the principle: *As you sow so shall you reap*. Animals are farmed and killed to eat on an industrial scale. Wild animals are hunted to extinction. The oceans are emptied of fish and ancient forests are clear felled, killing everything they contain until what was once a place of natural beauty is reduced to a hellish wasteland. Life is treated with distain while money, the instrument of greed, is worshipped. Driven by greed, corporations ravage the world. Driven by pride and power, governments instigate wars. Humanity kills until there is nothing left to kill.

Russell Brand was castigated for making light of the murder of tourists on a Tunisian beach. Nothing justifies the killing and our hearts go out to the families of those who lost their lives but Brand has highlighted the hypocrisy in our society when it comes to killing.

Humanity is at war with Life and our cruelty to defenseless creatures is inexcusable. Vivisection, even in the name of science, is abuse. Fruitarians and vegans but also vegetarians do their best to dissociate themselves from this bad karma but the rest of us are culpable for crimes against Life by our choice of consumption. We have only ourselves to blame for the consequent repercussions of this karma.

On Earth we are top of the food chain. Not so after we leave. Howard Storm and Ian McCormack experienced darkness and bitter cold. They were in a terrible place that had no definition in space and no sense of time. It was full of fear and brooding evil. Ian was aware of people around him in fear and confusion. His description reminded me of cattle milling around in the marshalling yard of an abattoir.

If there is no God but only Life then the laws of Life would apply to humans in continuous living. The operations of these laws are evident in the natural world around us. I was walking in a park

## Chapter 17 – Karma

when a picture of life in a pond caught my attention. The image of a dragon fly larva, looking like some demonic extraterrestrial, with a tadpole in its claw, left a lasting impression on me.

Many tadpoles hatch from the spawn but only a few make it to become frogs. Life is only concerned for the tadpoles that become frogs because the frogs perpetuate the cycle of Life. The tadpoles that don't make it to froghood are eaten. That is fine by Life. Everything is recycled, nothing is wasted. It is the law of the jungle, eat or be eaten!

The human race may be subject to the same inexorable law of Life. I believe if we prepare for continuous living by living out of love and arrange to be collected at transition into the extra-physical by one who is already connected into the Light then we may progress there into luminous beings ourselves; beings of light in the super-physical realms of the Universe. From there, I imagine we will seed our own incarnations in the physical womb of biological life; the equivalent of tadpoles that made it to become frogs. I am speculating but …*as below, so above* I see no reason why the Universal cycles of human life should be that different to the life cycle of frogs.

From the death experiences of Howard Storm and Ian McCormack it would seem that those that have not made the effort to reach for the polarity of light and have not lived out of love, they are harvested into the polarity of darkness, along with those who have neglected love by living lives of cruelty and violence or selfishness and greed. We may not like talk of being harvested but if that is what we do to other organisms why shouldn't it be done to us. Karma!

In the Indian traditions we are told we can counteract karma by seeking a satguru to ferry us across the ocean of maya. Maya is a Sanskrit word meaning the *illusion of forms*. We are warned by the Indian philosophy that if we are deluded by maya we will be lost.

The role of the satguru is to help us see through the illusion of materialism and come into the knowledge of who we truly are. The satguru warns of the danger of following the mind and provides a meditation practice to help us release ourselves from its grip. In Sanskrit *sat* means true and a guru is a teacher who leads us from darkness into light.

I know of two satgurus who have helped me personally on my journey through life. One is called Amma. Amma means *mother*. Amma travels the world giving people hugs. She comes to the Alexandra Palace in London every autumn where I go to receive a hug. Hugs from Amma can help us make the passage across the ocean of maya because as we hug her she connects us to the Life source of living light. Obviously we need to be living out of love and believe in continuous living but it is worth the effort of a humble hug to get connected.

The other is Prem Rawat also called Maharaji. He is very wise with a great sense of humour and you can feel the power in him. I don't know how anyone can live without his gift of knowledge. Everyone who receives the knowledge-of-self through him is connected into the light and harmony of the Universe, with the sweetness through the abiding presence of Life in the heart. If you want to cross the ocean of maya in a metaphorical cruise liner go to Maharaji.

Most people say they don't need a guru. That may be true for a few. They think they don't need someone to lead them from darkness into light. Howard Storm and Ian McCormack numbered amongst them before their experience of death. Not so now, they have both devoted their lives in service to Yeshua who led them both from the darkness into light and from death back to Life.

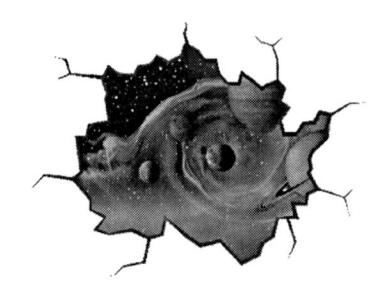

# Chapter 18

# Saved by the Light

While Howard Storm was lying beaten and broken a voice inside him, which he recognised as his own, called for him to pray. Initially he resisted because he despised religion, he didn't believe in God and he didn't know how to pray. However the voice persisted and he made a feeble attempt at prayer. Immediately the merciless mob, threatening more hurt and howling that there was no God and he was a coward to pray, began to withdraw.

As he lay in the dark Storm felt more pain in his body and emotions than anything he had felt in his life before. For him death was no relief from suffering. Seeking annihilation to escape physical pain he ended up in more bodily and emotional pain than he could ever have imagined.

Some will attempt to dismiss Storm's NDE as schizophrenia, or delusion induced by the attack of his gastric juices that were eating him alive and his rescue from abandonment as commencement of the surgical intervention to save his life.

Most of us will use our reason to explain away anything that threatens us in the comfort zone of our belief or disbelief. Howard Storm was in a zone of extreme discomfort. He didn't attempt to reason away his horrific experience. Approaching the ultimate pit of darkness and despair, he dropped his pride and began to pray.

Though prayer was diametrically opposed to everything he believed in, Howard Storm began to pray. In desperation and self recrimination he prayed for help. In his words,

*"How ironic it was ending up in the sewer of the universe with people who fed off the pain of others! I had had little genuine compassion for others. It dawned on me that I was not unlike these miserable creatures that had tormented me. Failing truly to love, they had been led into the outer darkness where their only desire was to inflict their inner torment*

*onto another...These debased people may have been successful in the world, but they had missed the most important thing of all, and now were reaping what they had sown."*

Despite the fact that Professor Storm was contemptuous toward religious people who prayed, believing they were just kidding themselves, he continued praying a mumbled jumble of the Star Spangled Banner, the Twenty-third psalm, with bits of the Lord's Prayer and the Pledge of Alliance to keep the terror at bay.

Then he recalled from his childhood the hymn, 'Jesus loves me'. He started singing it out loud which increased the fury of his persecutors but also their distance from him. At the same time the impenetrable darkness was broken by a single star which grew rapidly as he sang and then in an instant it exploded into a blinding light around him that incited the angry mob into a storm of derision and rage but also drove them back and away, deep into the darkness.

Albert Einstein once said the most powerful force in the Universe is the power of prayer. There is an example of this in Howard Storm's story. When he was a professor of art, one of the students who entered his faculty was a nun. He spoke to her and asked that she respect his atheism and not to speak of her religious beliefs in the class. She agreed but she and her congregation began to pray for him on a daily basis and continued to do so for thirteen years before his death experience. I believe that battery of prayer had a major part to play in way he was called to pray and the response to his prayer.

This could a good time to run a polarity check. If you are up for it, review the previous paragraphs and consider carefully the thoughts and emotions they brought up in you.

- What reaction did the mention of prayer and hymns, Jesus and the nun bring up in you?
- How did your reactions compare with those who attacked Howard Storm?
- If your thoughts and emotions were angry how did they make you feel?
- Did they make you feel happy and elated or unhappy and depressed?

# Chapter 18 – Saved by the Light

A feature of NDEs is reports of angels – not with wings and harps but immensely loving beings of light. You don't have to be a Christian and read the Bible or join the New Age with crystals, drums and feathers but you will need an open mind to accept how Howard was rescued!

Howard Storm has told his story to millions in books and talks to groups and on radio and television appearances, including the Discovery Channel and the Oprah Winfrey Show. Watch his *Focus* interview on the internet if you are reacting and doubt the sincerity of the man.

Both Howard Storm and Ian McCormack experienced telepathy in their NDEs. Prayer is telepathic communication. Telepathy works between the dimensions. The law of subsets allows extra-physical beings to pop thoughts into our minds.

If in your polarity check you reacted negatively to the mention of prayer, nuns, Jesus and angels, it might be worth considering where those reactions might have come from. Are you absolutely sure your thoughts always originate in your brain. Have you considered the possibility they might come from somewhere else?

We may find fault with religions. They tend to drag interpretations and outmoded fashions from a past generation into the present making it more difficult for the current teachers to get their message across. But the message is always the same. Contemporary teachers use the language and dress in the fashion of the day. Every generation has its own Life teachers and our generation is no exception.

In her mid-nineties, my mother was a fan of Russell Brand. She read his book and watched him on the internet and she said his message is the same as Jesus', and remarked he even looks like Jesus. She said Jesus would have dressed, behaved and spoken like him to the people of this generation and Brand is ridiculed and vilified by the establishment just as Jesus was in his day.

"Nothing ever changes," said my mum, "*Every generation makes the same mistakes.*"

# Continuous Living

The figure of a young man appeared in the light that came to Howard Storm. Were it to happen today, he might have been mistaken for Russell except for his height!

In Howard's words, *"Soon the light was upon me. I knew that while it was indescribably brilliant it wasn't just light. This was a living, luminous being about eight feet tall and surrounded by an oval of radiance. The brilliant intensity of the light penetrated my body. Ecstasy swept away the agony. Tangible hands and arms gently embraced me and lifted me up. I slowly rose up into the presence of the light and the torn pieces of my body miraculously healed before my eyes. All my wounds vanished and I became whole and well in the light. More importantly, the pain and despair were replaced with love. I had been lost and now was found. I had been dead and now was alive."*

Howard Storm received many teachings when he was taken by the luminous being into the polarity of light, before he was returned to his physical body in an operating room in Paris. He said he was rescued by Yeshua and Yeshua led the teaching and Howard Storm's life review. Yeshua said it doesn't matter what religion we belong to. What matters is that we love whoever we are with.

Howard Storm was taught that there are good people in bad religions and bad people in good religions. It is easy to judge religions but in doing so we are in danger of losing the essential message of love and tolerance they are supposed to carry. Many people, disillusioned by religion have turned to spirituality and when Yeshua spoke of religion he was including all paths that are centered on love and spirituality.

It is unfortunate that religions display intolerance. Intolerance reveals falsehood in religion and betrayal of everything religion is supposed to stand for. Intolerance has nothing to do with love. Jehovah Witnesses worship Jehovah but I have found them on the whole to be amongst the most tolerant, sincere and genuinely loving people in any religion I have ever come across. I haven't met a bad JW yet and yet they are despised.

It is the love people share not the religion they declare that counts. Howard Storm is now a Christian pastor yet in his interviews on the internet he radiates love and sincerity. There is no pretense in him. He lives the message of the one who saved

# Chapter 18 – Saved by the Light

him, to live out of love and love whoever we are with. It is clear from watching him he isn't preaching, he is just speaking from his heart. He is living from certain knowledge that Jesus saves. It makes me cringe whenever I read that on a poster outside a Christian church but then I haven't been lifted out of hell like Howard Storm or Ian McCormack. If I had their experience I might have a more tolerant attitude toward religion!

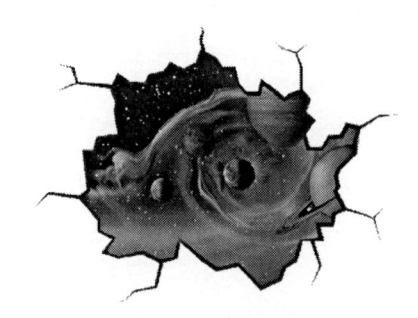

# Chapter 19

# Forgiveness

Ian McCormack wasn't religious. He said he was an atheist. He wasn't bad, he was just was a typical young man of his age, traveling the world, surfing, enjoying the occasional girl friend and living from day to day on savings and the occasional job.

While he was in the ambulance on his way to hospital, after being stung by Box jellyfish, pictures of his life started to come into his vision from when he was a small boy. He remembered he had read somewhere that sometimes people have a playback of their lives when they are dying.

He became desperate. He was too young to die but he realised he was dying. That is when he had a vision of his mother calling him to pray. She was thousands of miles away in New Zealand, but she had been woken in the night with a vision of her eldest son with bloodshot eyes and had been told by a voice in her head that he was dying. She was told she should pray for him so she had started to pray.

If we never pray and don't believe in God we may resist the idea of prayer having squirmed our way through a wedding or a funeral in church, with a vicar rattling away prayers that mean nothing to us, but Ian was desperate and he was dying. Like Howard Storm he didn't know how to pray. The poison was getting to his brain and it was beginning to close down. He couldn't formulate words.

Then he remembered his mother telling him to pray with his heart not his head. From his heart he begged the God he didn't believe in to help him pray. Immediately the Lords Prayer began to well up within him. When he came to the word, *forgive us as we forgive others* he begged for forgiveness.

That was when a voice inside his head called on him to forgive the Indian who had shoved him out of the taxi onto the street

## Chapter 19 – Forgiveness

and the Chinese men who refused to drive him to hospital. His immediate reaction was, 'You must be joking! But the prayer stopped. It wouldn't go any further.

He knew he had to forgive those men to continue the prayer so he let go of his longing to get back at them for their callous behaviour and forgave them. He then promised if ever he pulled through he would devote his life to the will of God whatever that may be. Peace descended on his heart and he knew deep down that whatever happened to him he was going to be OK.

When he got to the hospital and the doctor and a team of nurses got to work on him Ian's hopes rose but as their efforts failed and he felt himself slipping away, hope died with him. Then when he woke up alive and fully restored but in a terrible cavernous black space, presided over by foreboding evil with people telling him in his head he was in hell he despaired.

Ian was full of a deep sense of dread and desperately confused. He had prayed to God and had felt the peace of forgiveness in his heart so why was he in hell? He began to weep and call out to God in anguish, "Why am I here, I asked you for your forgiveness, why am I here, I turned my heart to you, why am I here?"

Then a brilliant white light pierced through the darkness from above him and shone in his face. He then felt a sense of weightlessness as he was lifted by the light and began to rise like a speck of dust in a sunbeam.

In Ian's own words,

*"As I looked up I could see I was being drawn into a large circular shaped opening far above me – a tunnel. I didn't want to look behind me in case I fell back into the darkness. I was very happy to be out of that darkness.*

*"Upon entering the tunnel I could see that the source of the light was emanating from the very end of the tunnel. It looked unspeakably bright, as if it was at the centre of the universe, the source of all light and all power. It was more brilliant than the sun, more radiant than any diamond, brighter than a laser beam. Yet you could look right into it.*

# Continuous Living

*"As I looked I was literally drawn to it, drawn like a moth into the presence of a flame. I was being pulled through the air at an amazing speed towards the end of the tunnel – towards the source of the light.*

*"As I was being translated through the air I could see successive waves of thicker intensity light break off the source and start traveling up the tunnel towards me. The first wave of light gave off an amazing warmth and comfort. It was as though the light wasn't just material in nature but was a 'living light' that transmitted an emotion. The light passed into me and filled me with a sense of love and acceptance. Half way down another wave of light passed into me. This light gave off a total and complete peace. I had looked for many years for 'peace of mind' but had found only fleeting moments of it. At school I had read from Keats to Shakespeare to try to find peace of mind. I had tried alcohol, I had tried education, I had tried sport, I had tried relationships with women, I had tried drugs, I tried everything I could think of to find peace and contentment in my life, and I'd never found it. Now from the top of my head to the base of my feet I found myself totally at peace.*

*"In the darkness I hadn't been able to see my hands in front of my face but now as I looked to my right and to my amazement there was my arm and hand – and I could see straight through them. I was transparent like spirit, only my body was full of the same light that was shining on me from the end of the tunnel. It was as if I was full of light.*

*"A third wave broke off from the main source of light as I neared the end of the tunnel. This wave hit me and as it did total joy went through my being...My mind couldn't even conceive where I was going, and my words couldn't communicate what I saw. I came out at the end of the tunnel and seemed to be standing upright before the source of the light and the power. My whole vision was taken up with the incredible light. It looked like white fire or a mountain of cut diamonds sparkling with the most indescribable brilliance...a voice spoke to me from the centre of the light. It was the same voice as I had heard earlier in the evening. The voice said, "Ian do you want to return?" I was shaken to learn there was someone in the centre of the light and whoever it was knew my name. It was as though the person could hear my inner thoughts as speech. I then thought to myself, "Return, return, to where?"*

*"Quickly looking behind me I could see the tunnel dissipating into darkness. I thought I must be back in my hospital bed dreaming and I closed my eyes. "Is this real, am I actually standing here, me, Ian*

## Chapter 19 – Forgiveness

*standing in real life here, is this real?" Then the voice spoke again, "Do you wish to return?"*

*"I replied, "If I am out of my body, I don't know where I am, I wish to return." The response was, "If you wish to return Ian, you must see in a new light."*

Ian's first thought was that the light was going to cast him back into the pit but to his amazement in response to that thought a wave of pure unconditional love flowed over him. It was the last thing he had expected. Instead of judgment he was being washed with pure love.

Ian then realised he was being asked to return to his physical body. He was told later that he had been forgiven when he had asked for forgiveness in the ambulance and had forgiven the men who had left him to die. He also discovered that he had been lifted out of hell by Yeshua the same luminous being that had lifted Howard Storm into the light.

When Howard Storm was asked to return to his physical body he argued because he didn't want to leave the love and the light but he was told he was not ready to remain in the light. Though he had been rescued from hell there was a lot he had to do before he would be ready to enter heaven. Ian also resisted because he didn't want to leave the place of love and light but he was shown his mother and his family and many others he could help by returning so, like Storm, he conceded. In the hospital mortuary fifteen to twenty minutes after he had been pronounced dead, to the shock of the staff preparing his body, Ian opened his eyes.

Like Howard Storm, and so many others who go through a profound NDE, subsequently Ian McCormack became deeply religious. In both cases, it was only because of prayer that the two men had been prompted to pray. Without prayer they would both have been lost.

We live in a perilous situation where prayer and forgiveness are out of fashion. It is easy to be angry and hateful when someone hurts us. It is hard to forgive and even harder to have the humility to ask for forgiveness.

As I write I am in my campervan. I have just put it back on the road after spending nearly £700 replacing the radiator, water pump, and cam belt after a nasty neighbour stuck a screwdriver through my radiator because I parked in front of his place. Why the hell didn't the guy put up a no-parking notice?

People told me to go to the police even though I have no evidence but I know he did it. I was thinking of driving over his fence in revenge but then I thought that wouldn't achieve anything. Instead I reasoned that maybe his action forced me to change the cam belt. It was five years old and if it had broken it would have been a lot more expensive, not to mention the hassle of a breakdown. So I sent him a telepathic thank you for saving my engine. It may seem silly but that was how I chose love and forgiveness instead of hate and spite. It was not easy and required an effort of reasoning on my part.

I was picking red currents with my daughter Josephine and talking about forgiveness. She reminded me of the importance of self forgiveness. She said all too often we beat ourselves up for things we have done or omitted to do. She said loving and forgiving ourselves is as important as loving and forgiving others.

She then spoke of how fortunate she was to be working in the NHS as a midwife because despite paperwork and cleaning up a lot of mess, there were amazing opportunities to share unconditional love and witness the wonders of new life emerging.

Then she spoke of the way she strokes the backs of the new borns as she hands them to their mothers. That reminded me of Howard Storm describing his back being stroked by Yeshua when he was lifted out of hell. It struck me that Yeshua had acted toward Ian and Howard somewhat like a midwife, delivering them out of darkness and into the light.

# Chapter 20

# Call the Midwife

When a woman is expecting a baby she contacts a midwife so she can have a safe delivery If she is too busy or can't be bothered or if she is proud and wants to do it alone she only has herself to blame if she loses her baby.

In the *Focus* interview on You Tube, Howard Storm said compared to the world he went to this world is like a dream. NDEs support this view and many spiritual teachers liken the physical world to a womb. To people who believe in continuous living what we call death is birth into the extra-physical. Judging from the experiences of Howard Storm and Ian McCormack, extra-physical birthing might be when we most need a midwife to secure a safe delivery.

Ian's situation was a bit like a baby dropped in the mud by an irresponsible teenage mother who didn't know she was pregnant and didn't care either. That is the attitude of so many people today who don't know that they have an extra-physical body and that they are responsible for its birth.

They believe death is the end, rather than the beginning and in their ignorance they eat, drink and be merry and mock the spiritual until it is too late. Like lemmings such people lead each other over the cliff called death and fall into an abyss of darkness as that was what they chose to believe in.

If we believe the Storm and McCormack return from death we might be wise to call on a *cosmic midwife* to help us birth into the extra-physical. Each of us is the parent of our extra-physical body. Each of us is responsible to contact a cosmic midwife to prepare for a safe delivery of our extra-physical self into the next phase of our continuous living.

So who are the cosmic midwives? They are people who are living or have lived an exemplary life, someone who has been an example of self sacrifice and unconditional love. It could be an

## Continuous Living

historical spiritual figure like Yeshua or his mother or Francis of Assisi or it could be an ancestor such as a spiritual grandparent or your cosmic midwife could be a living teacher like Amma or Maharaji. It is up to each of us. We have to call the midwife! The cosmic midwives are there for us all but like any other midwife, they will only come if they are called.

So how do you call up a cosmic midwife? If you want someone on-planet and like the sound of Maharaji go get knowledge, or if you prefer Amma, go get a hug. If you choose someone off-planet, like Yeshua, your guardian angel, or your grandmother, use telepathy! Take time. Find a quiet spot where you won't be disturbed. It may be in your bedroom or in your garden. A church is ideal but so is nature. Imagine the saint, angel or ancestor of your choice is standing or sitting beside you. Conjure them up in your mind. I promise you they will be there immediately you tune into them.

They will be right beside you in an invisible extra-physical body. You won't see them but you might feel them. Then speak to them as you would to anyone you know. They already know you so you can be as familiar as you like. A joke or two won't go amiss. It doesn't matter how you approach them. It only matters that you are comfortable and true to yourself.

Have a chat; let go of your sensible mind for a moment and be as a child talking to an invisible friend. When it feels right for you, ask from your heart for the cosmic midwife of your choice to be with you throughout your life on Earth and to be there for you at the time of your physical death. Also ask them to help you to love others unconditionally and to love and forgive people that annoy and irritate you. Midwives give mothers exercises. Loving and forgiving people unconditionally is the best exercise for cosmic birthing.

No mother would just call the midwife and leave it at that. She would develop a relationship with the midwife and maintain regular contact prior to the birth. So it is with your cosmic midwife; try to keep up regular contact.

I keep a shrine for my cosmic midwives to help me remember and maintain the connection. I don't worship them because they are men and women just like me but I burn a tea-light candle to

## Chapter 20 – Call the Midwife

signify my choice of polarity for the light. I sit with them and meditate for a few minutes when I can.

If I remember them for a few minutes in the day I find they are with me all day. I find it easier to meditate if I call on their company by lighting a candle because when I attempt to sit in silence and connect I am tormented by my cynical mind. The constant and invasive stream of thought is oh so sensible and so scientific but such a killer of joy.

Your shrine might be just a picture of your grandmother on the mantelpiece and your connection maybe just a candle lit by her picture from time to time. If you find it impossible to pray or to associate with any religious figure but feel comfortable with your Nan, so long as you know she chose the light in her fashion that should suffice. It is not what you do or how you do it that matters is is your intent for love and light that counts. That will not go unnoticed.

All we have to do to choose the light is to intend for the light. All we have to do to choose love is to live out of love. All we have to do to attain forgiveness is to forgive others. All we have to do to find a teacher is to be a teacher.

If you prefer to connect with the good God, that will work also because the relingo *God* is recognised by the cosmic midwives. If you chose a name your friends would respect it. The cosmic midwives are our friends so whatever name we choose they will respect. All the midwives in the Universal Health Service (UHS) work as a team. They are only concerned where our hearts are at, not our heads.

The need to do something positive to attain the polarity of light was highlighted in Ian McCormack's experience of death. It seems to me tragic that the other people in the cavern of darkness were left there. Why didn't they make a connection with someone decent that could guide them to a place of safety?

Reading Ian's account made me sad, not for him but the people left behind, all the people left in the dark because there was no one who cared sufficiently to pray for them. I want to put out a clarion call for all of us to be as cosmic midwives for each other. Yeshua was no different to us as a man but his heart bled for the suffering of humanity. If he was a freedom fighter, starting out

## Continuous Living

with anger and hatred for the Romans, at some point he must have seen the light and realised love not hatred is the answer. He became a healer and a teacher preaching love but then succumbed to anger in the Temple and paid for it with his life.

Yeshua taught that we should forgive not hate those who annoy us; even priests and bankers! Like Ian, it was his act of forgiveness when he was dying that enabled the light to lift him, after a descent into hell, into heaven. And just as Yeshua returned from death to help birth us into the light, so Howard, Ian and Eban were resurrected from death and now are acting as cosmic midwives, helping to deliver others out of the darkness and into the light. It was predicted that the dead would rise again. NDEs and ADEs are it!

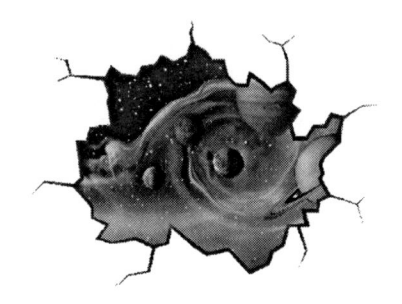

# Chapter 21

# My Experience of the Light

I had an experience of the light from Yeshua that filled me with the same love and joy as Ian McCormack and Howard Storm described. Fortunately I didn't have to die, but I did have to go to Medjugorje. Medjugorje is a village in former Yugoslavia, where Yeshua's mother was appearing on a regular basis to a group of young people.

Despite turning my back on the Catholic Church I have always had a good relationship with Yeshua and his mother who I feel to be my brother and his mother to be my spiritual sister. I have always felt their presence in my life even if I never saw them; that is until December 29, 1989 when I saw them both at Medjugorje.

We can have good relationship with Yeshua; even if we don't go to church or are not particularly good. When I saw him I had let down my wife and family and was divorced and I was sleeping with a woman out of wedlock – I married her later!

I was going through one of my anti-Christian phases because I was reading *The Mists of Avalon* a novel based on the Christian destruction of paganism in Britain. I was then and still am a believer in paganism and the Wicca and I hadn't been to church for years.

So there you have it, a lapsed Catholic, divorced, living in sin, advocate of paganism and lover of witches meets Jesus Christ. I think Yeshua thinks I am a bit of a joke really. I know whatever I do or say he still loves me because whenever I come back to him after forgetting him for years he never seems to be bothered. We just pick up like mates from where we left off. It was very much like that when I arrived in Medjugorje.

In the summer of 1989 I had met an American who was staying with a friend of mine in Sherborne, Dorset. He was on his way back to America after a visit to Medjugorje. He told me all about it and I was interested but had no intention of going there myself. However, a couple of months later when I visited my friend in Sherborne again the American was still there.

He was frantic. He said every attempt he made to return to the USA was thwarted and he had been told he wouldn't get away until he persuaded me to go to Medjugorje. He said he wanted to get home in time for Christmas so please would I go to Medjugorje before the end of the year.

Despite my anti-Catholicism I felt so sorry for the guy I agreed to go. Within days of my partner Hillary booking us on a tour to Medjugorje, his obstacles were removed and he was on his way home.

We landed at Dubrovnik airport and took a coach to Medjugorje. Hillary said her main interest in the trip was the drive along the Adriatic coast over mountains that reached down to the sea. She wasn't overly interested in religion either; more a tourist than a pilgrim. The run was spectacular and I was getting dream recalls all along the way. Scene after dramatic scene I recognised from dreams. I have prophetic dreams that warn of danger or forewarn of life changing events. My life was about to be changed!

As soon as we arrived in Medjugorje the tour operator said there was a service beginning shortly in the Church. We found the room designated to us in the village, unloaded our gear and set off to the church. When we went in the service was already in progress. We settled into a pew near the back.

I went into an immediate negative reaction. The priest was going on about the passion of Jesus. I had heard it all before at Westminster Cathedral Choir School where I had boarded as a boy and in practically every church I had been to since.

I really didn't want to be in church and was wondering why on earth I had even come to Medjugorje. Why had I caved into the pleas of the American? Bloody compassion! He could have got back to the USA. He was just coercing me. Scenes from the *Mist of Avalon* were running through my mind reminding me of the passion of the pagans at the hands of the hypocrite Catholics.

Then, in the midst of the storm of discontent in my head I found myself listening to the celebrant calling on us to meditate on the

## Chapter 21 – My Experience of the Light

scourging of Christ prior to his crucifixion. A wave of compassion went out from my heart to the brave young man Yeshua who had to endure agony for opposing the religion of his day! I sent up a little prayer to him, "Please take my negative thoughts away."

The response was immediate. It felt as though a hand went through my head driving away the thoughts as though they were a pack of monkeys that had taken up residence in my cranium. It was a very real experience. In that instant I recognised the true nature of negative thoughts. They had filled my head with blackness and as they were swept away by the slap of that hand they took the darkness with them. The space in my head was filled instead with lightness and relief, so much so that I burst into tears. That was when I saw her.

To my right the figure of a tall, graceful woman had appeared. She was looking at me. She didn't say anything and she wasn't clearly defined. She looked like one of the pictures of heavenly figures my children used to draw at their Rudolf Steiner School. I knew her by her presence; a presence I had known since my childhood. It was Mariam. Words came into my head immediately,

*"Please introduce me to your son."*

A young man then appeared directly in front of me. He was dressed in a white gown blowing in the wind. I could feel the heat of the wind and particles of desert dust were stinging my face. It was surreal. He was very clear. He stood looking at me. He felt and looked exactly like my brother Stephen. He was so very familiar that I laughed with joy at the sight of him again. It was my beloved Yeshua!

A beam of light came from his heart to my heart and immediately I was filled with indescribable joy. Every cell in my body was filled with light and singing with joy. It was tangible ecstasy that cannot be described; it has to be experienced to be believed.

The light was not like the light we see. It was more like a fluid. It was a liquid light that was sustaining and delightful. Just look at the word I have just written. It was invented to describe what I was experiencing and what Howard Storm and Ian McCormack had experienced before me. Delight filled us with joy. I experienced the ecstasy Howard Storm and Ian McCormack spoke of when they were lifted out of the darkness. It is hard to describe the energy of a living liquid light which is also love and peace and joy but when you do you will know the plasma that is the essence of Life.

The liquid light started to move up from my feet and down from my head toward my chest. It was a strange sensation, a normal forehead but nose still in bliss then normal nose but joyful chin. As the liquid moved through me it took the ecstasy with it. The two figures had gone. All I was aware of was the sensation of the liquid joyful light doing its own thing in my body.

The liquid light was a fluid as real as water or oil in our world. It was a real form of energy. I have no doubt I was experiencing it in my extra-physical body. Hillary was sitting right next to me. She didn't see or feel anything I was seeing but she could see I was going through some sort of experience because my face was full of light and joy, I was shaking and tears were pouring down my cheeks.

I felt the energy move down my arms as if it were water dropping down in two tubes. Eventually it all ended up in my hands. My hands were frozen and vibrating with the power in them. When I came out of ecstasy I knew what I had to do with it.

First I asked Hillary to take my wallet as I didn't want the taint of money on me. Ever since that day I have never felt comfortable with money. As soon as I get it I get rid of it by giving it away. Giving money purifies it by transforming it from an instrument of greed into a conduit of compassion.

I had brought a manuscript of *The Vortex Theory* with me to Medjugorje. I intended to ask a visionary to hold it up to Mariam to bless. I never got a chance to do that because as soon as I got to the village her son did it for me.

I asked Hillary to open the doors of the church so we could leave. I knew I mustn't touch anything. My hands were stuck out in front of me vibrating with power as we walked through the village. I couldn't do anything with them nor could I feel them; they were numb. Hillary opened the doors to the house where we were staying, unlocked the door to our room, opened my

# Chapter 21 – My Experience of the Light

suitcase, got out the manuscript and placed it in my hands. As soon as the manuscript touched me the power went into it with a woomph! My hands were left tingling and I could flex them again.

What I experienced was a real, tangible, scientific reality. It is no good people telling me I imagined it or I was hallucinating. If I spoke to someone who has had an NDE or a transcendent experience they would understand immediately.

It is hard to explain something that others have not experienced because of denial and sarcasm. That is why most individuals who have had an NDE don't speak about it. People always attempt to explain away things they have not experienced or don't understand. That is the nature of the mind and the basis of ignorance.

Before my rapture I experienced the reality of my brain. It is like a radio-recorder that had picked up programs that reasoned in my head. I could hear them in my head talking incessantly with their sequence of logic as though someone left a radio on. I had always identified with the thoughts in my head.

When a button was pushed by the priest going on about the passion of Jesus a program of reaction started. I thought it was me reacting until that moment when the thoughts, jabbering in my head, were sent flying. They had been living in my head like parasites. They were filling my brain with darkness. They were infecting me with nasty thoughts. They were running me, making me react in certain ways. I felt like shouting out something rude at the priest but I was in church and had been brought up as a boy by a strict mother not to run round, shout or be rude in church. But it wasn't me that wanted to shout out obscenities, it was the parasitic thoughts that possessed me.

With all due respect to Howard Storm and Ian McCormack after their return from death they became pastors but they tend to speak in a language that would incite a negative reaction in most people who don't relate to religion; the same sort of reaction as I experienced in the church of Medjugorje. Only religious or spiritual people will relate to what they say, so their experiences are in danger of being lost to people with anti-religious sentiments.

# Continuous Living

The experience I had at Medjugorje was not religious; it was real. A real form of energy came into me from Yeshua. It was a power that came through him but I already knew it was not unique to him because I had experienced it before. I had received the same power of liquid light and joy in Orlando in 1979, when I touched Maharaji in a ceremony called *Darshan*. It was so powerful my knees caved in and I had to be helped out of his presence and laid out on a mat to recover – along with many others! It was when I went to Medjugorje I realised it did not matter if people turned to Maharaji, Amma or Yeshua, they would be led from the same darkness into the same light!

The power is not the spiritual teacher, it comes through the teacher. It is a bit like mistaking the light bulb for the light. The bulb transforms electric power into light and transmits it. People like Yeshua, Amma and Maharaji are like light bulbs. They have reached a stage in their evolution where they can transmit immense amounts of plasmic power – which is a form of electricity – without burning out.

Not many people can handle that sort of energy. They tend to be rare. Yeshua let it be known to the people in his day that he was the only cosmic bulb blazing at that time when he said *"I am the way, the truth and the Life. There is no way to the Father but through me."* (John 14:6)

Access to that liquid light of Life is our birthright. We are growing our extra-physical body, in a way like building a house. At some point the house needs to be connected to the mains electricity so the lights can go on. We could do it ourselves but we might get electrocuted. It is safer to find a certified electrician to connect us and make sure we get the right amps for our fuse box. People like Yeshua, Maharaji and Amma are cosmic electricians. They make sure we only get the voltage we can cope with at any time. But once we are connected we shouldn't worship the electrician. We should just get on and let the light shine through us.

It is good to recommend the electrician who connects us to people who are still in the dark, but it is the electricity not the electrician that they actually need. The confusion comes from the fact we cannot access electricity without an electrician; unless we are stupid!

# Chapter 21 – My Experience of the Light

My problem is I think I am as good as any electrician. I was doing a wiring job for my third wife Liesbeth in Australia in 2005. I stuck a wire in my mouth to tear the plastic sheath from the copper with my teeth forgetting the plug was in the wall socket and switched on! Mr. Twit almost had an NDE!

Fortunately she was a doctor but she never let me do any electrics after that. It is the same with the spiritual sparks, I am good at pointing the way but I don't classify myself as competent to make the connections. Yeshua was the anointed one in his day. Maharaji and Amma are amongst the finest cosmic electricians coming out of India in our day.

There are others. Take your pick. It is up to you to choose who you want to plug you in. For Russell Brand it was the transcendental meditation of the late Yogi Maharishi that plugged him in so he could switch on his inner light

Some people I know connect through ayahuasca. It is amazing to think a plant could act as a guru and lead people from darkness into light. I suppose if people are anti-guru, anti-religion anti-spiritual and anti-everything else all you can do is throw them a vine to grab onto!

# Chapter 22

# Plug into the Power

I believe the plasmic energy Howard Storm, Ian McCormack and I experienced as liquid light comes from the level of super-physical energy. I makes sense to me that someone would have to attain to a level where they can handle the full power of super-energy plasma before they can enter that realm of reality. Howard Storm, Ian McCormack and Eben Alexander along with everyone else who has ever had a resuscitation experience seems to be offered the opportunity to return to the Earth to do more work to prepare for continuous living in the extra-physical and super-physical.

If people choose to live in light continuously and have been connected to the Universal mains but are not ready to receive the full power of the super-physical it would seem obvious that they would remain in the extra-physical realm for a period while they learn more about the power of love and potential of compassion and had an opportunity to purge negative thoughts and emotions that have possessed them during their lives on Earth.

I imagine there might be a level in the extra-physical polarity of light acting as quarantine before we move onto higher levels, somewhat like where people used to go to in the days of sailing ships to clear infections before entering Australia and America. When I was a Catholic I was told there was a place called purgatory where some people go to purge before entering heaven.

If you imagine an extra-physical being as a house plugged into the domestic mains, a super-physical being like Yeshua would be more like a city connected direct to the National Grid. Ian McCormack's description of Yeshua is the way we will all be one day if we choose the way of Love and Light and can leave Facebook for a minute to take the trouble to call up a cosmic electrician to get our extra-physical house connected to the Universal mains.

# Chapter 22 – Plug into the Power

If we are too distracted with our terrestrial trinkets or are too busy or can't be bothered or are too proud to plug into the power we could end up in perpetual darkness. Hell appears to be an extra-physical continuum designated for beings that have chosen not to be plugged into the Universal power. There may be nothing intrinsically wrong with them. They would be just the equivalent of houses that are in continual darkness because their owners never bothered to call up a cosmic electrician to connect them to the mains or they could be like houses that were once connected to the Universal mains but have somehow lost their connection. There is nothing intrinsically wrong with hell either. Julian of Norwich said hell is a deep expression of God's love. Howard Storm was told by Yeshua that he lifts people out of hell if they call on him to do so but it doesn't happen very often. A lot of people would prefer to be there in the dark with their mates than in the light with a whole bunch of spiritual people reminding them they were wrong!

If you look out for electric pylons you may see a line of small older pylons running close to a line of big modern ones. The little pylons were the original grid. They no longer run the main power. That is done by the new big pylons. The little pylons now just carry extra capacity. They were once the biggest and best power lines on the landscape now they are taking second place.

Maybe there was a time when the *Ancient of Days* was the brightest light in the firmament but since then newer and brighter lights have appeared to outshine him; most especially the ones coming up from the mud and mire of mankind like Yeshua.

He was once a maligned man hanging on a cross. Now he seems to be a light shining like a million stars, outshining the original luminaries by a billion lumens. Maybe the Ancient of Days is jealous of us. Maybe he did something to covet us when, as leader of the elohim, he was involved in our genetic engineering. Maybe he did something to our DNA to enable him to control us by possessing our minds. Maybe that was how he enslaved us!

Pylons will one day rust away but continuous living is continuous. It is not so easy to get rid of a cantankerous old spark bemoaning the good days of ancient yore when men were mere worms wriggling at the base of the evolutionary ladder and the

## Continuous Living

morning star shone down upon them from a great height smiling benignly on their singularly squirmy situation! I wonder if the legend of the great light Lucifer falling into darkness is the story of a super-energy being dropping in energy speed from that of super-energy to the level of extra-physical energy.

If a line of pylons is old and rusty, should one fall it might take other ones with it because they are all connected by lines. When the lines come down there is a blackout.

You could think of hell as the place where the rusty old pylons ended up in a heap. Other rubbish got dumped forming a junk yard where useless old things accumulate.

The thing about continuous living is that super-physical beings that have fallen into the extra-physical are continuously living plasmic fields. They never die; they just live in darkness because they are not connected to the source of liquid light. Hell, if you like, is the junk yard for Universal Life where disconnected plasmic life forms accumulate. You could think of it as the sewer under the super-physical, a realm in the extra-physical running with cosmic rats.

As I perceive it in the human situation we are given a body but we have to decide if we want to connect it to a source of super-physical energy so we can live continuously in light or leave it disconnected and live continuously in darkness.

All houses are in the default position of darkness until the electrician makes the connection. Of course the houses are illuminated while they are being built but once the builders have left with their lights, unless the electrician is called in to connect the house to its own supply of electricity, it will be left in perpetual darkness.

It seems logical to me that humans that have not decided to live in the light or bothered to live out of love cannot proceed toward the source of light in the super-physical because they will not be recognised and will be rejected.

If they haven't connected to love and light by spontabeously living out of love and light or by securing a connection through one who is already connected, they may still be alive in an extra-physical body but with no where for them to go and

## Chapter 22 – Plug into the Power

nothing for them to do in the darkness they would be lost to the light because all humans are destined for the light and designed to enter the super-physical realm where, sustained by plasmic energy, perceived as liquid light, they can grow into super-physical beings of light.

It is my speculation that plasmic entities from the darkness want us because they seek a continual supply of extra-physical Life to feed on. I think it is the electromagnetic emotional and mental energy in the human extra-physical body they want. I believe they invade our minds as thoughts to dissuade and distract us from the light. I reckon they collect people who are not recognised as being in the light.

Howard Storm and Ian McCormack were lucky because people prayed for them and they had the humility to pray. Many are not that fortunate because they neither pray for help nor have others pray for them.

# Chapter 23

# Earthbound

People often wonder what the purpose of human life is. I think the primary purpose is to decide whether we want to live continuously in the light or in the dark; in love or in fear. Whatever else we do with our lives on Earth is of secondary importance in my opinion. You may have a different opinion about the purpose of your life but when you shed your physical body nothing you have done will matter then apart from your acts of loving kindness and the care you have shared.

I have always had an inner certainty that we live continuously and death is not the end of Life. How we live now determines how we go on in continuous living. This is clear in NDEs where people meet a being of light and have a review of their lives. Nothing in their lives seems to matter apart from their acts of loving kindness. We may know this in our hearts to be true but we kid ourselves otherwise; or rather the thoughts we are so attached to kid us otherwise!

People tell me I am wrong to be concerned with the afterlife and the only thing that matters is life now. NDEs reveal the only thing that matters is that we live out of love now. If we share kindness and consideration for others in this life, Life will share kindness and consideration for us in the next. It is that simple. We don't need to be concerned about the afterlife if we live for the welfare of humanity, the Earth and all living creatures, giving of ourselves in selfless service to others. As well as NDEs there are other sources of information on continuous living. They don't have the same level of credibility as NDEs but they are worth consideration. Mostly they are psychic rather than religious.

One in particular is *Seven Steps to Eternity* by Stephen Turoff now published by Clairview. It describes how people that transition into the light can help collect other humans into the light when they are lost and confused after death, because they were not aware of the purpose of their lives while on Earth.

## Chapter 23 – Earthbound

It seems that people with open hearts and minds, even if they are not religious, can be collected into the light if they have love in their hearts. They in turn can then be trained to collect other people into the light.

Psychic sources suggest that people who make no provision for continuous living and don't realise they have died when they pass out of their bodies they may linger around familiar places on Earth thinking they still live here. This accounts for haunting. These people are earthbound. Extra-physical people who are earthbound are sometimes seen, sensed or heard. The relingo *ghost* is used to describe them.

The movie *The Others*, starring Nicole Kidman, gives a brilliant portrayal of earthbound people in the extra-physical; complete with the fog. It is one of her best performances, in my opinion.

Someone who is Earthbound becomes attached to a familiar place on Earth and they don't know how to move on. Unless they are collected and taken away to either polarity they could remain in the familiar place endlessly. When I was a child we lived near Bude on Stamford hill where a civil war battle was fought. At night sometimes footsteps could be heard crunching on the gravel outside. No one was ever there and my mother had a hard time keeping her nannies.

The Earthbound situation emphasises the need for a revision in our attitude to death and the irresponsibility of the materialist world view. If people were prepared for continuous living there would be far less suffering of the Earthbound.

**Earthbound people miss out on a world of incredible beauty. In the words of Dr Eban Alexander:** *"The worlds above are not general, not vague. They are deeply, piercingly alive... There are trees in the worlds above this one. There are fields and there are animals and people... When we ascend, in short everything's still there. Only it's more real. Less dense, yet at the same time more intense... The objects and landscapes and people and animals burst with life and color.*[1]*"*

---

1    **Alexander** Eban, *The Map of Heaven*, Piatkus 2014

# Chapter 24

# Journey into the Light

The study of NDEs makes it clear that ignorance of the purpose of human life is a major cause of loss and confusion at the end of our time on Earth. Preoccupied with our physical lives and our survival needs we often remain unaware that we are here to prepare for a journey into the Universe and miss out on the journey altogether. It is as though we are on a platform to get on a train but because we are ignorant of why we are there we miss the train.

People who are definite in their belief that death is oblivion are like those on the platform that firmly believe there are no trains or life beyond the platform. If you point to a train puffing at the station they will just look the other way.

There are numerous stories of bad people who choose the light on their deathbeds, beg for forgiveness and go into the polarity of light. They are like people on the platform who realise their mistake and leap onto a train just as it is pulling out and make the journey into the light.

There are good people who go into the dark because of their pride. Pride is the original sin because it can hold a person in a mindset that draws them into the polarity of darkness. Having told everyone there is no train they are not likely to jump on board when one pulls into the station.

Not permitting a last minute re-think, pride is the most common cause of rejection. Anyone can change their mind and choose to believe in continuous living and go for the light. No one is forced into darkness against their will. Everyone always has choice. It is just a matter of making it!

Original sin is the separation from Life caused by the grip which mind has on us. We can break the cycle of original sin by breaking the grip of the mind. Try doing it now. Close your eyes

## Chapter 24 – Journey into the Light

and imagine passing away into the darkness of death and oblivion. How do you feel?

Do you feel light and uplifted or dark and depressed? Now imagine passing away and stepping through a door into light and a new adventure in continuous living. How do you feel this time? We can break the grip of our minds by following our feelings rather than our thinking; intuition rather than reason. It is that simple. It is feelings that tell us what is right, not thinking!

To break the grip of the mind is simple but not easy. Meditation is the most powerful tool to help us break the grip. But we can practice breaking the grip by repeating the exercise of following feelings rather than thoughts. Try again! Imagine how you would feel if you chose to remain on a dark platform and ignore a train full of light as it pulls into the station. Then imagine how you would feel if instead you stepped out of the dark and through a carriage door into the light. All you have to do then is align with the feelings rather than the preconceptions.

Exercises like these to follow your feelings rather than your thinking can help you take decisions in your daily life because if death is a delusion our daily lives now are part of continuous living.

We can choose to journey into the light or choose the material option and live as though there is no Life beyond the physical plane and end up lost in darkness because we have been conditioned to believe in lies. To quote Dominique C. who wrote to me recently from France, *"During my childhood I was told that our western society knows it all on all subjects, and that everything else is a failure and absurdity, and what we have done as a civilization is "heaven" on Earth. Several decades later, I am in the full and painful impression that on most subjects we are conditioned to swallow lies...*

# Chapter 25

# God as Life

Love is the concern of the heart. Belief is the concern of the head. There are some who believe in God that post images of themselves on the internet, burning a pilot alive or decapitating a hostage. There are others who don't believe in God that help people in need and treat whoever they meet with loving kindness. Is it any wonder so many people have turned their back on religion and have become atheists?

We can be atheists. We don't have to believe in religion or the God of religion if we can appreciate the principle that the relingo *God* actually represents.

The clearest description of the principle God represents, that I have ever come across, is in a piece of channeling. You may not believe in channeling but I believe we are all channeling most of the time, only we call it thinking. I believe thoughts from the polarity of darkness are more invasive than those from the polarity of light because hell shouts in our heads whereas heaven whispers in our ears!

As I already suggested, we can tell which polarity thoughts come from by monitoring the feeling we have when we think them – or rather when they think through us!
- Do they make us feel happy or sad?
- Do they uplift us or do they depress us?
- Do they bring on anger or do they bring on joy?
- Most important, can we switch them off?

If thoughts invade our headspace and keep coming back when we try to dismiss them then they are definitely not good and we may need to make an effort to let go of them and that can be difficult.

If the channeling below conflicts with your disbelief in channeling, read it as a hypothesis and see how you feel as you

## Chapter 25 – God as Life

read it. I had to tell you it was a channeling because it comes from perhaps the most famous series of channelings in the world, *Conversations With God* by Neale Donald Walsch. The quote below comes from the end of the penultimate book, *Tomorrows God:*

*"Now I suggest that there is another word for God, the meaning of which you may all agree upon. When this word is exchanged for the word God, everything suddenly becomes simple and clear. Life is the one word in your many languages that comes closest to carrying the meaning that some of you are seeking to express when you use the word 'Allah' or 'God' or 'Brahman,' 'Vishnu,' or 'Shiva,' and that you hope describes or can be a container for 'the stuff that God is.'*

*"In one word, LIFE is the 'stuff that God is.'*

*"Life IS. Life is that which IS. It has no shape, no form, no gender. It has no color, no fragrance, no size. It is ALL shapes, all forms, all colors, fragrances, and sizes. It is both genders, and that which is genderless as well.*

*"It is the All and the Everything, and it is the No Thing from which Everything emerges.*

*"There is nothing Life creates that is not Life Itself. All that you see everywhere around you is Life, expressing. Life is everything. It runs in, as, and through everything. YOU are Life, expressing, and Life is you, expressing itself as you. Everybody else is Life expressing. There is no one, not a single living being, who is not an expression of Life. Even those you consider the worst among you are an expression of Life.*

*"None of these statements seems controversial. Not many people would seriously argue with them. The truth of these statements seems obvious on the surface.*

*"Now play a little trick on yourself. Play a little game. Exchange the word 'Life' for the word 'God' and see what happens. Watch you mind go crazy. Make the same statement, as above, but use the word 'God' in place of the word 'Life' and watch what your mind does with it."*

If we limit Life to biology then this extraordinary message makes no sense at all. However if we expand our horizons and allow for the possibility of a Living Universe then this remarkable piece of wisdom makes a lot of sense. It is easy then to treat God as the principle of Life if we see Life in terms of plasma physics rather than carbon base biology!

We can love Life, respect Life and care for Life and every living being. We are alive so to love, respect and care for ourselves is also to love Life. If we believe in the endless continuum of Life and embrace love in our life then as we pass out of this world, we can carry the momentum of a loving life on in an unbroken flow of continuous living and death we will not know because death will not be part of our reality.

It is our attitude of care for each other and every other living creature and the Earth that matters. How can religious faith in a God mean anything if those professing it treat women with distain and kill in God's name?

Treating God as Life removes the necessity for religion. This is wisdom for us all that is eternal and for everyone, religious people and non-religious alike. It is not who we are, it is who or what we believe and what we do that determines our destiny. Do we love killing or do we love Life. Do we love power or do we do all in our power to help others in need. Do we love money or do we love to share money with others knowing it is the living love of Life, not money, that we all need.

The beauty of physics is it opens our minds to the possibility of Universal Life. We can understand plasmic Life as Spirit and spiritual teachings in terms of resonance and frequency, attraction and repulsion, positive and negative. Instead of turning science against the eternal message of Life, we can employ science to make sense of that message and rewrite it in a language appropriate to our day.

Once we believe in Life as a Universal principle that is not restricted to biological life as we know it on this planet the entities called God in the Bible, the elohim of the ancient Hebrews, we can treat as expressions of Life, extraterrestrial maybe but still expressions of the same Life as in you and me. We don't have to worship them nor do we have to believe the record of them in the Bible as *The Word of God* because the Bible tells us precisely what the Word of God is, and it is not the Bible!

# Chapter 26

# God as Energy

The Bible reveals the Word of God as Energy. It also defines God as energy. This fits with God being Life as Life is a form of energy. Life has to be energy because everything is energy!

In the Bible the first verse of the Gospel of John reads:

*In the beginning was the word,*

*And the Word was with God,*

*And the Word was God.*

Word is sound. Sound is vibration. Vibration is energy so the first verse of John's gospel could be translated as:

*In the beginning was the Energy,*

*And the Energy was with God,*

*And the Energy was God.*

Religious people may argue against the definition of God as Energy but the Bible stands as it is written. The Bible is subject to translation and interpretation and there is no denying that this interpretation will appeal to scientists because the similarities between the descriptions of God in religion and the descriptions of Energy in science are remarkable.

- God is neither created nor destroyed:
- Energy is neither created nor destroyed,
- God is everywhere: Energy is everywhere,
- God is in everything: Energy is everything,
- God is all powerful: Energy is all power,

People talk of God as love and light. That fits with God being energy because love and light are forms of Energy. If Energy is not limited by the speed of light to physical reality then Energy could be the beginning and the end; the alpha and the omega.

We limit all energy to the level of Energy we perceive because we live in it. A problem with humanity is our tendency to limit reality to our perception of reality. We limit Life to carbon based biology and Energy to the physical speed of light. As soon as we lift our vision beyond the horizon we can see the principle we call God is a living, multidimensional continuum of Energy.

Treating God as Energy is pantheism. Pantheism is the belief that God is everything and everything is God. Pantheism is an appealing alternative to atheism or monotheism. Pantheism suggests God is the Universe.

I think of God as the Consciousness underlying everything. If particles of Energy are not acts of material they could be acts of consciousness. Thought is an act of consciousness. Particles of Energy could be more thoughts than things. Thoughts are particles of intelligence. If particles of Energy are thoughts then particles of Energy would be particles of intelligence.

A body of thought is a mind. If particles of energy are thoughts then the Universe could be a mind. Mind cannot exist without underlying conscious awareness. That brings us back to the quantum idea that consciousness is the bedrock of reality. One Consciousness holding an intelligent, Living Universe in being is what I believe in!

As soon as we drop the material philosophy of Democritus and embrace Quantum theory instead everything fits, everything makes sense. All it took was the vortex theory to explain away material substance and replace it with energy in spin.

The use of *Word* in John's verse is seminal as it implies intelligence. It fits with the Hermetic idea that reason underlies everything. Words, written and spoken, convey reason and presuppose intelligence. Spoken words communicate thoughts and written words hold the memories of reason; they store intelligent information. The *Word* implies that Life is the innate intelligence underlying Energy.

Noise is sound, it is vibration, it is energy but it does not convey intelligent information apart from informing us a lorry has

## Chapter 26 – God as Energy

arrived. The *Word* in John's verse represents the principle of reason that wills things into existence. In the original Greek, the *Word* was *Logos* which means *Reason.*

Hermes taught that the Logos wills things into existence and John reiterated that by asserting that through the Word or Logos everything was made. He also related the Word to the light that shines in the darkness and the Life that brings light into the world.

The most evident thing that brings things into existence is Life. It is the intelligent Energy of Life that brings great symphonies and works of art into being. But Life represents the polarities of both light and dark. Works of evil as well as good come from Life manifest in mankind and reason can support the polarity of hatred and darkness, cruelty and anger, pornography and the promotion of addiction as well as the polarity of goodness.

The living Energy of Life works for us all if we know how to work with it. It is playful and yet at the same time it is awesomely powerful. If we trust it enough anything is possible but the trust has to be total! In January 1997 I set out on a journey around the world with nothing but a £10 note. I laughed at the gods because of the ridiculousness of my situation. I told them they would have to look after my needs as I was unable!

With total trust in the providence of Life I had booked a flight to Australia and organised a lecture tour despite the fact I had no money for the air fare. I used the £10 to buy a one day London Transport travel pass for £4, gave £1 to a beggar at Waterloo Station as my 10% tithe and pocketing the fiver set out on my travels without a care in the world.

I had absolute confidence my needs would be met. By the end of the day my air fare was paid as advance royalty on a book I was writing, plus £20 for my pocket. I then stayed on a houseboat at Teddington Lock next to a Rolls Royce in which I was driven to Oxford the following day.

I met a dancer who kept me going on love, lychees, cigars and cans of Guinness until I managed to make it to Australia where I met a beautiful doctor who later became my wife and after that we made a fortune.

# Continuous Living

Meanwhile I went on to Bangkok where I met a *Mufti el-Allah*. He said Allah had told him to look after me so he gave me $4000. Then I went on into Cambodia where I stayed with a friend of a friend who had married the President's daughter. I saw the sights pillion on a BMW motor bike, one of four outriders escorting the Mercedes on family outings. Allah provides. Thanks to Allah I managed to visit Australia three times that year and I did it all on faith and faith has provided for me ever since.

If God is Energy and Energy is everything then as forms of Energy we must come from God and depend on God for everything. If God is Life we must come from Life and depend on Life for everything. Use whatever word works for you. It is the Universal principle these three words represent that counts, not the words. Only the fickle human mind make an issue of words.

What struck me about the NDE descriptions of a surgeon living for the welfare of others against a surfer living only for himself was the abundance of Life experienced by the surgeon against the poverty experienced by the surfer in the absence of Life.

NDEs suggest that those who live for others attain the abundance of Life in continuous living whereas those who live only for themselves end up with nothing in continuous living.

It all boils down to faith or fear. If we live in faith that Life will provide for our needs, naturally we will share with others from what we receive. If we live in fear of Life we will take from others to satisfy our greed. To those who have faith to give, everything is given but from those who have fear and take everything, it would appear that everything is taken away from them.

# Chapter 27

# Faith or Fear

It was an overcast day in July 1969 when I was fishing off a surfboard about half a mile out from Crackington Haven on the North Atlantic coast of Cornwall. I had a line of feathers attached to one end of baler twine, threaded through the belt loops of my cut away jeans.

Four Pollack and two Sea Bass were trailing in the water behind me, strung along the other end of the twine. Suddenly there was a resounding slap. I looked round and saw a ring on the water about two metres across. I thought it must have been an enormous fish that jumped.

It was. A dorsal fin was slicing through the ocean in ever decreasing circles around my board. I was studying zoology so I identified the three metre shadow in the water immediately. Shark! To begin with I wasn't worried because there was no record of a shark attack in Cornwall.

Then I thought 'what if the shark went for my fish and the baler twine got caught in its teeth, it could drag me off with it'. So I pulled the fish out of the water and onto the board. Now the thing about Sea Bass is they have razor sharp spikes on their dorsal fins.

Flopping about on my little board they started to lacerate my legs. Blood was running into the sea. Knowing sharks have evolved to go into a feeding frenzy at the scent of blood, I panicked. Fear seized me. I paddled for all I was worth toward the bay and the safety of the shore. My frantic splashing must have scared the shark away because I wasn't attacked but I couldn't shake off the fear of attack. Once I gave way to fear it held me like a vice. I experienced the expression *gripped by fear*. Fear devoured me, not the shark.

I had always been fearful. I didn't have much self confidence and I would worry a lot about little things, but that was the most intense experience of fear I had ever had. Seven years later I received the gift of knowledge-of-self and began to meditate.

What I noticed was my life didn't change much, if anything I had more reason to be anxious as I was married and had no proper income. Anna and I had toddler twins, Josephine and Jessica, Rebecca was a baby and Sam was on his way. We were living from hand to mouth in a renovated shack, but I wasn't anxious. A quiet confidence resided inside me. Instead of being in a state of fear I was in a state of faith.

Faith had replaced my fear. It wasn't an intellectual faith. It had nothing to do with belief. It didn't come from my mind. It was a certainty that everything is OK. Life had been, was and would always be OK for me. It is not that I don't worry from time to time, or ever get stressed, but anxiety never clings to me like it used to. It comes, hangs around for a bit and then slips away because fear no longer gets a grip on me. As I came to know who I truly am, faith replaced fear.

The knowledge didn't stop me being stupid. I messed up and lost Anna and then Hillary but thankfully they had received knowledge too and it made a big difference to the outcome of the divorces. Though there was short term anger and recrimination there was no long term acrimony. The certainty of knowledge and trust that everything is working for the best has allowed a lasting friendship. All three of us have found that is possible because in our hearts fear has been replaced by faith.

I still think a lot but the grip of my mind is gone. There is a presence in me I now know which is stronger than my mind and gives me a sense of assurance that nothing can ever harm me. It is a deep certainty I cannot describe. It is a knowing I will always get by; I will survive come what may. My faith does not come from anything or anyone outside of me; it is a faith that comes with knowledge of the Life in me.

Faith and fear are polarities. Fear is built into all animals because it is essential for survival. Faith is something we have to work for. Fear brings up anger. Faith brings out love. That is why I am so grateful for the gift of knowledge-of-self that connects me to the source of Life within me.

# Chapter 27 – Faith or Fear

I didn't have to work for faith or struggle to release fear. It was more an indulgence in something delightful I had discovered – actually my parents discovered it. I resisted it to begin with. My mind found all sorts of objections but when I eventually followed the preparation – now called the Peace Education Program – and I accepted the gift of Knowledge my polarity switched from fear of Life to faith in Life.

Most people associate the word *faith* with religious affiliation but people who have had a personal experience of connection through Yeshua or another true spiritual teacher will appreciate what I am talking about. Faith is more a state of being than belief. It is realised in the heart not the head. Faith comes with the commitment to know, love and serve Life so we can live continuously with Life, sustained by the liquid light of love.

In India there is a greeting that goes *Sat Chit Anand*. Sat means truth, Chit means consciousness and Anand refers to the bliss and also love. Yeshua is also known in the Sanskrit as *Sananda*. *San* means saint, holy or health. Sananda means *the holy one who brings bliss, healing and love*.

It doesn't matter which tap we go to, to get a drink of water or what name we give the tap. It is the water we need not the tap. The other thing we need is the thirst to make us drink the water.

If you are inclined to religion, go to the tap marked 'God' and let Yeshua refresh you. If you are disinclined to religion then go to the tap marked 'Life' and let Maharaji provide for you. These are the taps I recommend because I have drunk from them both and can assure you it is the same sweet water that flows through them. You might prefer the tap marked Amma. She is very sweet.

If you prefer Russell Brand's TM tap go for that. If you find another tap drink from it because the taps are all connected to the same mains supply guaranteed clean and pure with the certainty of quenching thirst. However, if you are thirsty and drink from a muddy puddle when there is a good tap nearby don't blame me if you end up sick. We each of us have to take responsibility to make informed decisions and check the science as well as the religion before we believe anything.

## Continuous Living

My parents converted to Catholicism whereas I was born into it. I didn't have an opportunity to choose for myself when the water was poured on my head. All I did was howl. As far as I am aware infants belong to the light by default as do innocents with mental retardation.

The mind begins to grip at the age of reason which is seven. Jesuits know this to be true which is why they say, *"Give me a child up to the age of seven then you can have him for the rest of his life."* That attitude is abuse because before seven brain templates of conditioning form which are hard to reverse. I think children should not be programmed into belief or disbelief but allowed to make their own informed choice as they grow up with the capacity to reason and exercise their free will.

There are many ways to connect to the light. I know the manager of a sheltered housing community of elderly people, many of whom struggle to cope with maintaining their independence. The level of care he gives is amazing. He is always there for everyone in his community, always smiling, never cross or irritable. He doesn't belong to any religion and is not on any spiritual path he just lives out of love. He says he has his own way. He doesn't need to follow anyone else's.

What is interesting is that he wears a gold crucifix around his neck on a gold chain. Not because he is a Catholic or a Christian but because he likes to wear it. Some may say it is superstition but it is not. Any talisman is a way of identifying polarity. By wearing his crucifix he maintains a tangible link with Yeshua without prayer or meditation, religion or the Bible. He connects with Yeshua simply by choosing to wear a symbol of him.

We don't have to do a lot. We just have to make it clear which side of the fence we are on so the Universe is clear about our intentions, and icons speak intentions. You may feel drawn to put an icon in your wallet or handbag.

If you wear a medal or a picture of a spiritual teacher around your neck it doesn't have to be on show. It is no different to carrying a picture of a loved one. It is just a way of keeping in touch. It has nothing to do with the world at large. It is good to be discrete and avoid embarrassment because your mind and the minds of others will resist any gesture you may make toward the

## Chapter 27 – Faith or Fear

light. I once wore a picture of an Indian guru round my neck. I don't even know his name. A friend put it round my neck. I just loved wearing it! It doesn't matter what totem we use if it represents our chosen polarity.

I have a little book on St Francis of Assisi I was given. I feel his presence whenever I dip into it. I also set up statues of Buddha in gardens and around the house. Quan Yin porcelains make beautiful decorations on a mantle piece. I recently set up a statue of The Lady of Grace in a memorial garden to my mother. I light a tea-light candle in front of her and by that action I state my intention. Lighting a candle indicates choice of the light. This is why the lighting of a candle in a shrine is significant.

The evil God of the Old Testament commanded in the Bible that icons were to be destroyed, altars and oak groves torn down. It began with Abraham desecrating the altars at Mecca and continued with the smashing of statues and stained glass windows at the time of the Reformation when Bibles in the vernacular were printed. Not realising the evil God was trying to disconnect people from their chosen path toward the light, people reading the Bible thought they were doing the right thing vandalizing sacred places.

What people who read the Bible don't realise is that maybe this move toward monotheism was part of the coup by one of the elohim to cut off people's connections to the light. It could have been part of his plan to get everyone to worship him. Even today this continues with dynamiting Buddhas in Afghanistan and ISIL, in the name of the God of Abraham, desecrating sacred places in Syria and Iraq along with killing followers of Zoroaster who believe in Mazda – which means the light!

Like a great pincer movement with materialism on one side and monotheism on the other, the Lord of Darkness, symbolized by the all seeing eye, is crushing spirituality and making it harder for people to connect with the polarity of light. In the nineteenth century Charles Darwin proposed his theory of evolution which provided materialists with an opportunity to engage with the monotheists, and connecting through contention, they funnel the unwitting into Mordor!

# Chapter 28

# Intelligent Evolution

If Consciousness is fundamental to Life and energy, intelligent evolution makes sense. When Max Planck, the originator of quantum theory said, *"I regard consciousness as fundamental. I regard matter as a derivative of consciousness,"* he set a trend amongst physicists to consider consciousness as fundamental to quantum reality. Perhaps it is time for biologists to get past the nineteenth century and catch up with the physicists! Consciousness implies intelligence. Intelligence cannot be excluded from evolution if consciousness is fundamental to everything.

Materialists believe that life on earth originated over billions of years by pure blind chance. They speculate that in a primordial soup, random atomic reactions formed amino acids which led to proteins, along with DNA and RNA and ATP and that happenchance led to Life. A long series of fortunate accidents led to the genetic sequencing of DNA and formation of proteins into membranes and enzymes which then evolved into living organisms by sheer luck!

But many scientists argue that the four billion years since the origin of Earth is insufficient time for the probability of this to occur by blind chance. The evolution of life from the accumulation of incremental random events suggested by Richard Dawkins, in his book *The Blind Watchmaker*[1] is just a fanciful idea. Even in inorganic life where probabilities are greater throughout the Universe than is available to organic life limited to the Earth, both in time and diversity of environments

---

1   **Dawkins** Richard, *The Blind Watchmaker*, Penguin, 1988

## Chapter 28 – Intelligent Evolution

in the galaxies, even then the evolution of life by pure blind chance it is highly unlikely!

Random mistakes as an account for the evolution of organic life is unlikely firstly because DNA is remarkably stable, and mistakes in copying are rare and usually disadvantageous. The idea that mistakes in the replication of DNA, through natural selection, could have led to the remarkable diversity of life on Earth is untenable.

Then there is the protein-RNA dilemma. RNA production requires protein. Protein production requires RNA. Compound that with the DNA dilemma. DNA requires enzymes and enzymes require DNA. Add to that the ATP dilemma. Enzymes are needed to form adenosine triphosphate but adenose triphosphate is needed to provide energy to form enzymes.

The probability of proteins (enzymes are proteins), ATP, RNA and DNA forming all at once and acting all together by chance is too low to be scientifically plausible. To quote the microbiologist Radu Popa:

*"The complexity of the mechanisms required for the functioning of a living cell is so large that a simultaneous emergence by chance seems impossible."*[2]

The evidence defies the hypothesis. Darwin's *tree of life* theory, that organic life started from a single source, is simply not the case. Life appears to have originated from numerous start points. It is more plausible to imagine that inorganic extra-terrestrial Life has influenced the evolution of organic life, perhaps through DNA resonance.

Maybe the DNA double helix has been modeled on double helix vortex strands that evolved in plasmic matter. Organic life on Earth could be the end point of an evolutionary process that has been going on throughout the Universe in countless galaxies over enormous periods of time.

2   **Popa Radu,** Between Necessity and Probability: Searching for the *Definition and Origin of Life*, 2004

# Continuous Living

Despite Richard Dawkin's best efforts to convince us there is no intelligence underlying evolution, just luck, many are not convinced. Many scientists who support evolutionary theory consider there must be some sort of intelligence involved in the evolutionary process but understandably they are not willing to accept the creationist view that God did it in seven days!

When the relingo *God* is replaced with the word *Life* the religious nonsense is removed. Most scientists would accept that if Life is established as a Universal principle then it could be responsible for the evolution of life on Earth. They may not know for sure how Life did it but if the process did start from random quantum interactions, because quantum theory allows for consciousness, intelligence would be implicit.

Ralph Waldo Emerson said, *Shallow men believe in luck. Strong men believe in cause and effect."* I believe in intelligent evolution driven by cause and effect not by luck or happen-chance. My contribution to the debate is to suggest that the causality of intelligence is in the shape of energy.

Intelligence is carried in patterns. Words and symbols are shapes. Pictures are shapes and patterns. Electromagnetic waves are shapes. The modulations of those waves that carry radio and television programmes are just patterns on shapes – modulated frequencies on carrier waves.

Intelligence depends on the transmission and storage of information and particles of energy lend themselves to that. The wave train is the active form of energy that can transmit information and the vortex is the static form of energy that can store information. Beyond the speed of light, waves of super-energy could be thoughts and vortices could be memories in the Universe acting as a mind. It is often said that thoughts travel faster than light!

Fundamental particles of energy appear to be memories. *The Vortex Theory*[3] provides a simple and straightforward account for

3   **David Ash**, The Vortex Theory. Kima Global Publishers 2015

## Chapter 28 – Intelligent Evolution

high energy physics by treating the two fundamental forms of energy as memories. Wave and vortex forms of energy, acting as memories but interacting randomly, according to the laws of physics, could provide an account for intelligence in terms of quantum interactions. Acting according to the consequence of their dynamic shapes, as wave train or vortex, particles of energy are governed by quantum laws of motion determined by their respective states of active or static inertia.

This allows for intelligence systems in the Universe to be described in terms of wave-particle mechanics. Governed by quantum theory, kinetics and Newton's laws of motion applied at the quantum level, the Universal evolution of Life could be viewed from a perspective of quantum interactions that lead, eventually, to ever more complex fields, structures and systems.

Rupert Sheldrake introduced the concept of systems intelligence in his revolutionary idea of morphic resonance for evolution but as he says:

*"Very little is known or can ever be known about the details of evolution in the past. Nor is evolution readily observable in the present...With such scanty direct evidence, and with so little possibility of experimental test, any interpretation of the mechanism of evolution is bound to be speculative."*[4]

In *The Intelligent Universe*, Sir Fred Hoyle reasons that evolution is driven by intelligence. Hoyle considered it a vast unlikelihood that life could have evolved from non-living matter without intelligence. In his words,

*"...it is apparent that the origin of life is overwhelmingly a matter of arrangement, of ordering quite common atoms into very special structures and sequences. Whereas we learn in physics that non-living processes tend to destroy order intelligent control is particularly effective at producing order out of chaos. You might even say that*

---

4 **Sheldrake** R, *A New Science of Life*, Paladin Books, 1987

*intelligence shows itself most effectively in arranging things, exactly what the origin of life requires."*[5]

The appearance of particles of energy, as memories, supports the premise that the Universe is a mind. As far as I am concerned the physical Universe, as we observe it, is the memory in the Universal mind. The Universe acting as a mind would enable learning from every event that occurs within it. There is no reason why evolution should not be a creative learning process within a Universal mind because the process of *intelligent evolution* is evident in our daily lives.

We put out ideas and see if they work. If they don't we adapt them until they do, or drop them and try something different. Evolving theories are subjected to criticism so that only the best survive. But all the time we are in the process. The work would never happen without our minds learning from the process of trail and error.

Consciousness and intelligence are never excluded from the human creative process. As we take rejection in our stride and learn from our mistakes we modify and make choices. There are lucky breaks but for the most part it is a learning process taken step by meticulous step. *As it is below so it is above, as it is above, so it is below;* why would it be different for a Living Universe?

When Howard Storm went through a life review, learning from mistakes was how his personal journey was presented to him by his teachers. We are all subject to the same evolutionary laws in our life progress. Why should it be any different for the Universe? If the Universe is a mind, it is obvious intelligence would underlie everything – including evolution – and a thoughtful Universe could learn through random interactions and events.

Life could be the culmination of universal intelligence evolving over aeons. Life screams intelligence. We are inundated by television programmes revealing the intelligence of plants and

5 **Hoyle** Fred, *The Intelligent Universe*, Michael Joseph, 1983

## Chapter 28 – Intelligent Evolution

animals. Slime mould smart as a Japanese engineer and apes solving mathematical solutions faster than humans, suggests it might be human arrogance that excludes intelligence from evolution.

There is no doubt Darwin's theory of evolution is brilliant and neo-Darwinian thinkers like Richard Dawkins have made contributions but perhaps they could allow for intelligence to be incorporated into evolutionary theory as part of their own evolutionary process.

Evolution makes sense as a means whereby Universal creative intelligence continually develops the multitude of forms that populate Life. A Living Universe would never stop evolving, learning and growing, breaking things down and then building them up again better than before. The extraordinary diversity and dazzling beauty we witness in Life doesn't call for an external Creator that made everything perfect in the beginning when it appears to be a system that is endlessly self creating through correcting imperfections.

Natural selection reveals mistakes. It is never easy for an author to spot errors. We rely on editors and critics to point out the flaws in our work. Life is harsher. If anyone is not up to par they end up as lunch. Random events are essential to creative evolution because they enable freedom. Creativity goes with freedom not repression

I believe Einstein made a mistake when he said *"God doesn't play dice."* It is through dice that Life plays. Chance is an essential ingredient in intelligent evolution; but it is not blind! The Western mind fails to see the underlying order in apparent chaos and the perfect operation of chance. Not so the Chinese mind. In the words of Carl Jung:

*"The Chinese mind, as I see it at work in the I Ching, seems to be exclusively preoccupied with the chance aspect of events. What we call*

*coincidence seems to be the chief concern of this peculiar mind and what we worship as causality passes almost unnoticed."*[6]

For intelligent systems, random events and unforeseen problems create opportunities for creativity and originality. Evolution may be witness to quantum levels of intelligence in biological systems building things by trial and error, watching them perform then breaking them down to reconstruct in a better way. This is the eternal cycle of creation, preservation and destruction – birth, life, death and then rebirth, as depicted in the mysticism of India in the trinity of Creator, Preserver and Destroyer.

The quantum process is a cumulative effect of the parts on the whole. Each particle of energy, treated as a particle of intelligence, could be a memory in quantum reality, contributing to the overall design of the Universe. In the chaos of atomic interactions, molecules arise as memories attesting to success over failure as time operates the remorseless process of selection. Perhaps heaven and hell are just a part of natural selection, operating between the planes of energy, in a Living Universe.

In biological life, organisms prey on each other. Many species avoid extermination in natural selection by evolving into parasites. The intelligence of living organisms can be remarkable in parasites. Often the nastiest are the smartest! Why would this be limited to biological evolution?

The mixture of polarities in humanity could predispose the planet as an evolutionary hot spot, where the polarities of love and hate, faith and fear, selflessness and selfishness work against each other in a selection process culminating in the segregations Howard Storm and Ian McCormack experienced as heaven and hell.

To protect themselves from invasive organisms, living things develop systems of immunity, which include isolation. Maybe hell is a device of Life to contain the contagion of pride,

6     **Wilhelm** Richard, *I Ching*, Routledge & Kegan Paul, 1951

## Chapter 28 – Intelligent Evolution

selfishness and violence. Our planet floating silently in space, with organisms of limited lifespan, is in effective quarantine. However, beyond the constraints of physical space, time and gravity, some protection against people who behave like pathogens may be an imperative. This would not be a God in judgment but more an immune system evolved by Life to protect the super-physical levels of continuous living from electromagnetic frequency contamination.

Matter carries memories. Light and sound carry information. If patterns we witness in the physical apply in the extra-physical, frequencies in super-energy could be what we experience as thought and vortices in extra-physical field matrices could be our experience of memory. Species could be thought of as Life memories storing the cumulation of evolution. The arrangement of cells in multi-cellular organisms may represent increasingly complex structures where ever more information can be stored, expressed and assimilated. Allowing for the existence of energy beyond the speed of light we open up the possibility for our own continuous living in unimaginably advanced systems of Life.

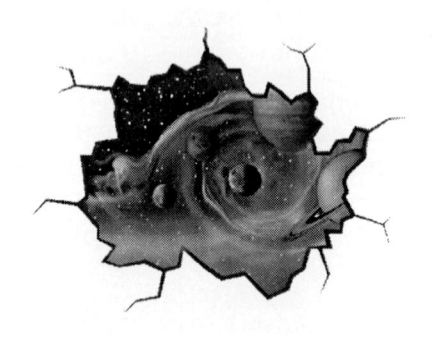

# Chapter 29

# Personal Evolution

If we accept the idea of continuous living and appreciate the importance of choosing our way in it then the first step is to open our hearts and our minds to new possibilities. That will let the light in; we need to see in a new light.

Seeing things is a new light is a bit like throwing open the curtains of a dark room to let the light in. Allowing new ideas into our heads is like opening the windows after a long hard winter to let in the soft fresh air, the birdsong and the sweet smell of spring time.

Facing the possibility of hell is the equivalent in our heads of putting a hard broom to the room so that nasty things that have been hiding in the dark can be swept into the light and banished into the wind through the open window.

Writing this book has had a profound effect on me. I have noticed things I was attached to in my head have fallen away and all my values and priorities have come into sharp focus. The after death experiences of Howard Storm and Ian McCormack have made me think again (the relingo *repent* literally means to re think) about my life and my destiny in continuous living and how much I have taken for granted.

Reading NDE accounts about people going into a tunnel of light then meeting a being of immense light and love followed by a life review and teachings before being returned to Earth never had the same impact on me. They were not life changing. But reading about the experiences of hell prior to the more normal NDE sequence of events, really rattled me.

It made me realise how addicted I am to certain patterns of thought which don't make me feel good and could be a distraction from what really matters. So many things I think about, I can do nothing about anyway, whereas there is a lot I can do if I attend to my personal evolution; if I take stock of how

## Chapter 29 – Personal Evolution

much I have taken for granted and how much more I could do with what time I have left on the Earth.

My personal evolution goes much like the evolution of Life. Things flow along a plateau of attitudes and behaviour for a long while, until something happens that brings about a change. Evolution is like climbing steps rather than going up a ramp. Change is not continuous. Change comes in shocks and this book has been a shock for me. I have written many books in my life but I have never before been kicked in the butt by one of my books.

I don't want to engender fear. That is not my intention but I do want to pass on the shock I have experienced by realising that hell could exist and could actually happen for me if I don't address selfish patterns of behaviour in my own life especially in regards to the selfish way I treated my first two wives. I am fortunate to have had the opportunity of caring for a centenarian before he died and then my mother before she passed away. Care work has taught me how to be selfless, patient and kind. I had to put my own life on hold for five years to focus on my mum but looking back those years were happy and fulfilling for me and my heart told me there was nothing I could do better as a son than to take care of my mum.

Heaven and hell exist side by side in our world. We only have to switch on the news or read a newspaper to be aware of that. But I have always thought of hell as being limited to this reality. I have assumed that if I don't get it right in this life I will reincarnate in another and have another chance so it doesn't really matter how I think or behave this time round. Realising it may not be me that reincarnates but rather the Life in me that comes back again and again has made me think again.

Realising I could be rejected by Life and end up in the sewer of the Universe, as Howard Storm so aptly put it, has made me review my life before I go through a life review. More than anything else I want to communicate my discoveries and realisations so others may rethink and reassess their priorities.

We all get so caught up in our daily lives we rarely think of death and are even less likely to consider the possibility of continuous living, let alone the impact of our present life upon it. We are

## Continuous Living

here for a purpose but the purpose we follow is all too often everything and anything but that!

It is so sad. All we can see is scaffolding. We are so preoccupied with the scaffolding we don't notice the beautiful house being built inside it. When the scaffolding falls away at the point we call death we mourn the loss of the scaffolding instead of rejoicing for a beautiful house that should be ours. Sadly because of our neglect is has been stolen by rats and robbers and they have destroyed it. We have no where else to go so we have to live on in all but a ruin.

We are here to learn, yes, but it is pointless if we learn lots and then end up in perpetual darkness because of our learning. We need to review to what extent we are learning facts rather than learning how to forgive.

Life is evolving through you and through me. We are both on the same journey. Our Life experiences are similar, along with our feelings and reactions, our needs and our desires. I could not bear to go into the light and leave you in the dark even though I don't know you.

It doesn't matter who you are or who I am because we are same being in different bodies. We are the same in our hearts. We both long for light and we long for love and we want to know that our future is secure for ourselves and our loved ones. We are all human and though we may appear different we are all so very much the same in the way we think and feel.

I have experienced the liquid light that is also love and joy and I can tell you, if you have not yet experienced it, it is worth more than all the money, all the possessions, all the success and recognition, all the relationships and all the places on this Earth. Nothing compares with it because it is the source of joy and peace from within. It is the inner source of happiness and fulfillment. It is the delight of the heart and living continuously would be a total loss without it.

When we have established a connection with the plasmic energy of Life, however we perceive that to be, then there is no fear or anxiety anymore. Life is lighter and more joyful. It feels more playful and it becomes hard to understand why everyone is so stressed.

# Chapter 29 – Personal Evolution

When you have the Life connection and are not stressed and depressed it becomes clear that people are stressed, anxious and fearful because they haven't yet established the all important connection with Life. Making that connection, to form a personal relationship with Life in the core of our being, is a very personal step of evolution for each and everyone in his or her lifetime. We may be alive, but we need to realise Life.

# Chapter 30

# Taking Responsibility

This book is all about each of us taking responsibility for our personal situation in terms of the possibility of continuous living. I have relayed NDE evidence that points to that possibility and supported it with my vortex theory. I have then gone on to consider implications and options. It is up to you whether you go along with my arguments or dismiss them.

Do you accept the materialist view that there is no continuous living and you have no source or do you prefer the view of the second century Gnostic mystic, Valentius, that there is continuous living but something is at work determined that we remain ignorant of that fact and unaware of our source. It is up to you to decide whether to believe the mystic or the materialist or someone like me who is attempting to bridge the divide between mysticism and materialism?

Ultimately you have to decide who and what to believe. Obviously to make an informed decision you need evidence but that may be hard to come by. People gain evidence of extra-physical reality through NDEs, transcendental and religious experiences, and psychic awareness. Some people explore their inner reality through meditation whilst others take psycho-active substances.

The life body would seem to have an inbuilt propensity for telepathy, extra-sensory perception, super-intelligence, super-normal powers and the experience of liquid light and it could be these faculties are suppressed by biochemicals in the brain.

Experiences with psychoactive substances like DMT (dimethyltryptamine) or LSD are usually explained away as a hallucination but it is also possible the brain may suppress extra-sensory perception. It could be that the chemical filters in the brain can be circumvented with psychoactive chemicals enabling the extra-physical level of Life to be experienced.

# Chapter 30 – Taking Responsibility

Many people who take psychoactive drugs experience one or other polarity and in some cases they experience the liquid light connection of Life. I believe this supports the model that psychoactive substances overcome the dull-down biochemical mechanisms in the brain to enable an experience to occur rather than being the cause of that experience.

While the prospect of exploring the extra-physical realms, by the chemical manipulation of neurological biochemistry, may seem attractive they can be addictive, destructive and a distraction from our true purpose which is to live out of love unconditionally and discover our source naturally. Russell Brand recommends meditation as he tried to break through with drugs and alcohol without success whereas he has succeeded with meditation.

I do not recommend the use of psychoactive substances to experience the extra-physical reality because psychoactive substances, including alcohol, can be dangerous. I believe the brain suppresses extra-physical faculties to protect us. Exploring the extra-physical while we are on the physical plane can exposes us to extra-physical infestation because the lowest energy levels of the extra-physical are closest to the physical. Alcohol and other recreational drugs could lift the protective brain filters, which might expose our extra-physical bodies to extra-physical psychic pathogens.

The relingo *aura* is commonly used amongst psychics. Psychics are people who are sensitive to extra-physical energies. The aura is a way they perceive the extra-physical energy field. My father was very aware of the aura in his healing work. Psychics sometimes see the aura in colour and technologies such as Kirlian photography provide scientific ways to study extra-physical electromagnetic fields. It is possible now to photograph the aura and study it.

Psychics see psychic entities attaching themselves to the aura. Extra-physical parasites would seem to attach themselves to our extra-physical bodies just as physical parasites attach to our physical bodies. Ancient wisdom teachings warn that they attach to us so they can feed on our life force, especially throughour emotions.

## Continuous Living

Young people who take recreational drugs could be likened to darling young things cavorting in the lush green grass of springtime, blissfully unaware that a herd of deer were there earlier dropping ticks. Alcoholics could be likened to seekers of comfort going to a place of good cheer and settling into armchairs unaware they are crawling with lice hidden in the seams. There are plenty of psychic parasites awaiting the unwary who abuse themselves with alcohol and drugs.

Human beings are vulnerable in physical life because the physical is part of the super-physical. All speeds between zero and the speed of light in the extra-physical realm is part of the extra-physical.

We may not see them but they can see us! We are exposed to all levels of reality in super-energy. By the law of subsets we are vulnerable to extra-physical infestation unless we are protected. Excuse the relingo; we are spiritual beings here to have a physical experience, not physical beings here to have a spiritual experience. Spiritual curiosity can be a distraction.

Atomic matter gives us a density which is not easy for plasmic beings to manipulate but they can influence us through our minds because the mind is in the extra-physical. The biochemical suppressors filter out extraneous extra-physical energies to an extent. Lift the filters and we are totally exposed.

Some plants with psychoactive properties can be teachers, especially if they contain DMT, but my recommendation would be to work with them only under guidance of experienced shamans who are properly trained in a native tradition and only if accompanied with prayer and ceremony to maintain a protected space. I do not think it wise to use teacher plants indiscriminately. They can help us become aware of a greater reality but they are powerful and can do harm as well as good, in the wrong hands or if they are used irresponsibly.

Books and teachers can also expose us to psychic attack. If they lead us into negative trains of thought they can open our minds to fear through which psychic entities can gain entry into our auras. Creating and maintaining sacred spaces is a way we can protect ourselves, along with prayer and meditation and sacred ceremonies. Praying, that is calling telepathically on the

## Chapter 30 – Taking Responsibility

protection of super-physical beings like Yeshua, is a way to keep negative extra-physical entities at bay.

Establishing a sacred space is fundamental in all spiritual traditions. The spiritual science of the sacred is based on the principle of energy subsets. In the sacred space, a super-physical being from a high level of super-energy is invoked to protect the space on the lowest energy level – the physical – from interference by extra-physical entities from the intermediate levels.

This principle is employed in exorcisms. Followers of Yeshua invoke him to use his super-physical power to banish extra-physical entities. Yeshua exorcised my head of parasitic extra-physical entities prior to his appearing to me in Medjugorje.

Unbeknown to me they had taken up residence in my extra-physical body. This act of purification of my extra-physical body would be in the physical the equivalent of being dewormed! Just as an animal needs regular worming so we need regular prayer and spiritual practice to purge us repeatedly of extra-physical infestation.

The problem with parasites is that after they are cleared, we pick up another lot. The same applies to extra-physical or psychic parasitic infestation. Maintaining prayer and meditation and sacred spaces is a way we can take responsibility for ourselves. Some people become psychotic after recreational drug use partly because of damage to the brain filters and partly due to psychic parasitic infestation the equivalent of a tapeworm in our reality, which is difficult to dislodge unless the appropriate remedy is employed.

Any involvement in psychic phenomena should be treated with respect and only undertaken if there is prior training in protection or there is supervision by people who are experienced in the field. It is all a matter of taking responsibility. Once we become aware of psychic and realities it is important to approach them with respect.

Psychic or extra-physical parasites, having lost their own connection to the living light of Life seek to access it through human beings. They prey on our mental and emotional energy while we are alive, if we allow them. Those they manage to capture after physical death, they would not literally eat but rather feed from their Life force and their emotional energy.

I was chatting with my nephew Ruben who is a surfer. We are on the same wave when it comes to the Living Universe. Ruben stressed the importance of approaching spiritual issues through science and the importance of getting out of our minds and into our feelings. The more we are in touch with our feelings the more integrated we become with our extra-physical life body.

There are many ancient sources of wisdom teachings that have been lost or suppressed by monotheism and derided by materialism but if we allow ourselves to go beyond the bounds of religiosity and scientism we can find many sources of spiritual advancement in antiquity. I recommend the book on yogic philosophy and oriental occultism that led me to my vortex theory.[1]

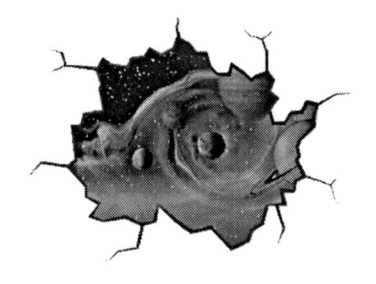

**Ramacharaka** Yogi, *An Advanced Course in Yogi Philosophy and Oriental Occultism* 1904 (Cosimo facsimile 2007)

# Chapter 31

# Mirror Symmetry

The words live and evil are mirror symmetrical. They represent the polarities of good and bad, living and dying, love and hatred, compassion and cruelty, faith and fear, joy and sorrow, light and dark, anger and acceptance.

In *The Vortex Theory* I predicted that the Universe is divided into two polarities represented by matter and anti-matter where the antimatter is both within and beyond the matter. The polarities of matter and anti-matter are mirror symmetrical and the interaction between them is the cause of gravity and the accelerating expansion of the Universe.

*As it is above, so it is below: As it is below so it is above.* I predict a polarity based on mirror symmetry in the realms of *super energy* beyond the speed of light, which fits with the other Hermetic principle that polarity exists at every level of the Universe. Mirror symmetry in super-energy is another way to account for heaven and hell.

I live in a small community that includes a doctor and his wife – she dropped *Descent into Death* into my lap – and a fine lady of ninety called Anne McEwen who founded the International Essene network. I have a lot of respect for her and what she says as an elder; especially as I believe Yeshua along with his cousin John the Baptist were educated in the Essene community at Qumran and trained as Essene rabbis.

Anne tottered in one day with a magnificent leather bound tome. For ninety she is usually sprightly on her pins but that day she was weighed down by Dante's *Vision of Hell*. Her daughter Helen is Buddhist and has been told about the existence of hell from that tradition. At the time of Dante there were no ideological exchanges between Buddhism and Catholicism.

If two totally disparate cultures described the same thing – like hell with seven levels – I believe it is reasonable to take the subject seriously. Because heaven is purported to have seven levels too, I am inclined to think that in heaven and hell we might be dealing with mirror symmetry. Maybe in super-energy beyond the speed of light, heaven and hell are states that are mirror images of each other.

I am of the opinion that heaven is not in the level of super-physical energy but is a series of seven sub-division levels in the space-time continuum of extra-physical energy which are sustained by the living, liquid light of life from the super-physical level.

I believe the power of the living light is stepped down through each level, much as voltages of electric power are stepped down by transformers, to enable people to survive in power ranges they can tolerate. I contend the mirror symmetrical polarity of heaven could be a domain in the extra-physical level of reality for beings to live in the light until they can ascend into the super-physical.

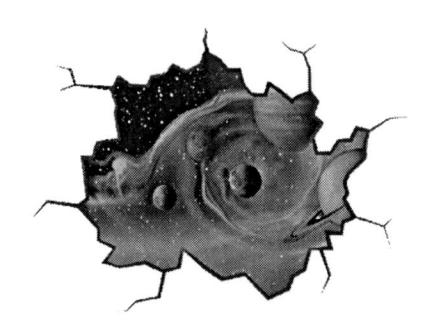

# Chapter 32

# The Polarities

By an act of will we can reverse polarity at any time but I believe the build up of predominant choices are likely to determine the polarity for each of us at the time when we shed the blindfold of physicality and wake up 'sorted' according to the consequence of a lifetime of choices or lack of them. Extra-physical sorting could be thought of somewhat like the settling of sediments in sedimentary layers. The layers could correspond to the levels in heaven and hell described by Dante.

I don't go along with the religious interpretations of Dante. My interpretation is that each person has free will and reward or punishment for acting on it would be an affront to that free will. The Howard Storm NDE suggests people who choose the polarity of dark continue living in darkness in the company of other beings that have chosen to exist in that polarity.

Rather than punishment or judgment, it would appear to be their choice. Many prefer to be in darkness than light. One thing that struck me in the Howard Storm story was the way his adversaries couldn't stand the light when it appeared in their midst. They shrank back into the darkness.

When he switched polarity and was rescued, every member of the mob surrounding him had the same opportunity to pray and escape the darkness in order to go into the light. The fact that none did so suggests that they were comfortable in the polarity of their choice and had no inclination to leave. I have a non-judgmental attitude to both polarities. They both exist therefore they must both have a function

I believe as each person makes their choices the extra-physical body stores the experiences of a lifetime rather like a memory stick. All experiences, good and bad, are of value to Life in the process of Universal evolution. During physical existence all polarities are mixed but in the extra-physical they separate out according to their predisposition following their life choices or lack of them.

As extra-physical bodies are electromagnetic they could be acting like electric charges. I perceive the separation of individuals, after each incarnation, into opposite polarities may be like building up charges in electric circuits to increase power and potential. I like to imagine it somewhat like a battery cell.

As incarnations on the physical plane pass through the gate we call death, they would, according to this premise, accumulate at one polarity or another. The buildup of live charges could create a tension between the poles that increases the power of the Life cell. Perhaps beings in each polarity hold consciousness so that the total spectrum between darkness and light is available for the experience of all conscious Life in the Universe. There is so much we don't understand about the Universe but patterns we observe in science could be reflections of patterns in the Universe: *As below, so above...* and Hermes did emphasise that polarity applies at every level in the Universe.

Electricity depends on potential gradients which are held by opposite polarities of electric charge. As each human is attracted to the positive or the negative maybe they contribute to the charge of that polarity and to the electric potential of all Life that exists everywhere.

Iron filings, once they start moving toward the pole of a magnet accelerate and become increasingly bound. In like manner once a person starts moving toward a preferred polarity they may become increasingly attracted toward it and more tightly bound by it. Those who choose love may progress to ever greater heights in heaven. For those who choose hatred progress may be to ever greater depths of hell. I believe these arcane descriptions were attempts to describe attraction and repulsion laws we now appreciate in physics.

Anne was talking to me about the Buddhist wheel of Life where a spirit incarnates through a number of lifetimes, to clear karma, as many individuals. Anne said that the experience of each polarity is not forever; extra-physical forms in the Universe are transient. Rudolf Steiner taught, from his occult knowledge, that each spirit incarnates individuals bodies much like organisms in

## Chapter 32 – The Polarities

a species. All dandelions are similar. They characterise their species. They are entirely different to roses but within the species of rose, all roses are similar. **Thus it is with spirits incarnating individual human souls.**

A lot of store is placed by the chemistry of genetics but we have yet to discover the impact resonance has on human personalities. I consider the physical body to be acting more like a blindfold, setting up tests, trials and learning opportunities for the soul or psyche being operating through its many incarnations.

In Eastern philosophy it is taught that each us chooses our parental and environmental situation best suited to balance our karma of previous incarnations. The way life unfolds, in either wealth or poverty, the influences of schools and churches, mentors and universities, not to mention friends and family, partners and work mates influence our being at every level.

We all battle with our emotions from time to time, most commonly with anger or jealousy. There is no need for us to feel guilt or shame about feeling these emotions. Experiencing these emotions is not bad because all too often the tendency to anger or jealousy is hereditary. It is what we choose to do with these emotions when they rise up inside us that matters. Do we release them on others and make their life hell or do we choose to reign ourselves in so as to act out of love toward them. Our thoughts and emotions represent challenges to overcome and opportunities to exercise choice of polarity. Love is usually the harder choice but that leads to spiritual growth toward the super-physical

I believe that as an incarnation we each have the potential to become a new super-physical being that is born with free will so that in this process we are trained and tested in our ability to choose. Mostly people react to situations habitually or instinctively.

Passing through physical death, people polarize to the light or the dark and contribute to the charge of Life. Occasionally people choose to act differently. Sometimes people exercise their free will. That supports their evolution into a new sovereign super-physical being.

## Continuous Living

The physical body has evolved within the biological sphere with the survival instincts and fears of animals that predispose it to selfishness. Locked into physicality by damping biochemical mechanisms in the brain, naturally the individual thinks it is a physical being and tends to behave like a biological organism where everything feeds off everything else. To act selflessly in parenting and in the pack – the family or tribe – is instinctive. To act selflessly toward entities outside the family, tribe or nation is not instinctive. That requires an act of will that runs against animal instinct.

It is unconditional acts of kindness and compassion toward other beings where there is no affinity of species or family, nationality or fraternity, church or community and forgiving those who persecute when the animal instinct it to retaliate and defend that the Universe could be selecting for.

Physical life in an animal body might be the best situation to test the ability of a being to choose to be different, to rise above normality, to ascend into nobility, to put the whole before the part, to change the course of history and the destiny of the Universe by transforming their own internal reality.

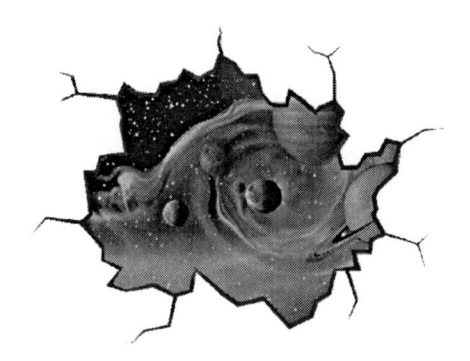

# Chapter 33

# Ascension and Healing

I believe we can become a super-physical being in the super-physical realm, capable of taking a creative role in the Universe with the capacity to seed our own stream of incarnations for learning and development. The process of transition from the physical or extra-physical into the super-physical levels of reality I call *ascension.*

I define ascension as a process in which the speed of energy, in a body of matter – be it atomic or non-atomic, physical or extra-physical – is accelerated taking the body from one level in the Universe to another. In *The Vortex Theory* I used the process of ascension to account for quantum tunneling. The acceleration of energy, in the vortices and waves of matter and light, would involve a quantum leap in the intrinsic speed of energy. Using the *...as below so above* symmetry, the ascension process in the macrocosm would be expected to emulate that in the microcosm.

I have reason to believe the speed of energy in the extra-physical to be in the order of twice the speed of light and in the super-physical to be in the order of sixteen times the speed of light. I cannot substantiate these figures scientifically but they do suggest the transitions between the levels of energy in the Universe are non-linear. The leap in speed of super-energy between the extra-physical and super-physical is considerably greater than that of energy between the physical and the extra-physical. I believe the extra-physical realms of reality to be associated with the planet Earth and the super-physical realms to be more Universal. Ascension into the super-physical would be release into the Universe.

Between 1992 and 1995 I was involved in the spread of a prediction of ascension throughout the world because through my vortex theory I saw the possibility of accelerating the intrinsic speed of energy in the vortices and waves of the

## Continuous Living

physical body to bring about ascension. At the time I spoke of the extra-physical as the hyper-physical.

I now consider it more feasible for the extra-physical, plasmic body to undergo ascension than the biological atomic body. Nonetheless, the vortex theory does allow for ascension of a physical body.

In the process we call death, the physical body expires and the extra-physical is released from its grip, so to speak, but the energy in the extra-physical would have been at the extra-physical speed all the time. The normal transition at physical death would be just a separation between physical and extra-physical, not an ascension.

If ascension occurs on the physical plane whilst the individual is in a physical body, the acceleration in energy speed could occur in the vortices and waves of the physical body as well as the extra-physical. The physical body would disappear out of physical space and time. This could account for the resurrection of Yeshua where his body disappeared out of the tomb. Reversing the process to descend the speed of energy in the body would then have enabled him to appear in the garden to Mary Magdalene.

Many people believe Yeshua went through a near death experience, descending first into hell and then rising to heaven – much like Howard Storm and Ian McCormack – before waking up in his body in the tomb and being rescued by his disciples. I can accept that account but my physics provides a way of accounting for more traditional beliefs in the resurrection and ascension of Yeshua that have been held by millions of his followers for millennia.

Nonetheless I am dubious about Gospel truth! For example I cannot accept that Pontius Pilate as a Roman prelate would have allowed Yeshua to be proclaimed King of the Jews by an INRI notice on his cross. Yeshua was a common man with no claim to kingship in Roman eyes. If he was crucified it would have been a standard crucifixion.

In AD 37 Antigonus II Mattathias, the last in the Jewish line of Hasmonean Kings, was captured in Jerusalem by Herod and Marc Antony and taken to Antioch to be executed. According to the Roman historian Dio Cassus he was scourged and then crucified, a

## Chapter 33 – Ascension and Healing

punishment no other king had suffered at the hands of the Romans. The INRI mockery is more the style of a Roman general like Marc Antony than a Roman prelate like Pontius Pilate.

I believe that the crucifixion of Antigonus inflamed the Zealot movement that led eventually to the destruction of Jerusalem just over a century later. I reckon the story of Antigonus was borrowed into the gospel stories when they were written centuries later.

However, the stories of miraculous healings never cease to intrigue me. My father, a medical doctor who practiced from Harley Street was a healer with extraordinary powers. I learnt a lot about the power of healing and witnessed it myself at the hands of my father. I am certain that some people have the ability to channel the liquid light of Light into the extra-physical body of an afflicted person and stimulate the electromagnetic template to bring about a regeneration of the physical body.

The word miracle is used through a lack of understanding about what is going on but I believe this modality of healing, when properly understood, will be standard practice in medicine in the future. I see no reason why Yeshua should not have been born with the ability to heal.

My father worked with another outstanding plasmic healer called Harry Edwards and they both experienced the Energy coming through them to bring about healings.

In the Howard Storm account, Yeshua healed him with his hands in much the same way as my father and Harry Edwards used their hands for healing. My grandson Ezra has inherited the gift. He can actually see the plasmic energy, as a violet light when he heals with his hands. Ezra started healing when he was five.

The power of healing could have been used to restore Yeshua to life after his crucifixion. The NDEs of Howard Storm, Eben Alexander and Ian McCormack witnessed recoveries that defied medical science. I believe all three of them died but were restored to life by the healing power of the liquid light of Life channeled by Yeshua. I believe it is that same power that brings about the ascension. The relingo for this power is *grace*. Indeed it is amazing grace!

# Chapter 34

# Understanding Ascension

The ascension opportunity is all about participating in an evolutionary step in Universal Consciousness. To begin with we are like living charged particles. The difference between us and subatomic charged particles is that they have a predetermined charge whereas we are capable of choosing the polarity of our charge and our destiny in continuous living. Through our experiences in the random unfolding of events on Earth, I believe collectively we are building up the wisdom and experience of Life management systems of the Universe.

I like to think of our collective history and individual lives somewhat like programs on a computer that can be re-run for management training so that consequences of choices can be reviewed. I believe each super-physical being as an individualisation of Universal Consciousness, has their own bank of incarnations in diverse cultures, in both male and female bodies, to provide opportunities for experiential review in the eternal *now*.

The extra-physical life bodies, after their time on Earth, in the mirror-symmetrical polarities, traditionally called heaven and hell, could allow consequences of choice, or lack of it, to be consolidated for the collective learning of all conscious Life. I think of this bank of experience somewhat like data accumulation on a computer hard drive.

The physical body could be viewed as a self-replicating, biological robot operated by a neurological computer, with sensory receptors and appendages for locomotion and manipulation of the environment. The driver of the machine is the super-physical entity operating through the extra-physical Life-field. The relingos, spirit-self, higher-self, angel-self, ascended master etc. would refer to the super-physical beings that seed consciousness and Life into an individualisation of universal mind, a psyche with the capacity to reincarnate in

## Chapter 34 – Understanding Ascension

physical and extra-physical bodies, a life-field described by the relingo, soul.

Your extra-physical body could be a plasmic field growing alongside the physical body capable of supporting its morphology. It could be collecting experiential data for you as a mind or a soul. This would be happening through physical and the extra-physical body and circumstances that you chose to incarnate in. Your 'soul' could be a store of that data for you in a domain – like a section in the hard drive of a computer – in the extra-physical level of reality. It would then be available for you to access from your super-physical level of existence, through the principle of subsets after you ascend

But you can have a brand new computer if you want. If the extra-physical body you are in now ascends to the super-physical, rather than ending up just another folder of data stored in your old computer, it can become a brand new living computer capable of storing its own files of data.

There would then be two of you! To comprehend that you need to appreciate that you are not your bodies in the physical, extra-physical or super-physical. These are just the means of individualizing you when in reality you are the Consciousness in all the Life systems of the Universe. You are the Consciousness underlying everything simply because Consciousness cannot be divided, it can only simulate division through the quantum of energy existing as a thought which can aggregate with other thoughts. These aggregations we experience as bodies and minds. This allows the individualisation of Consciousness. Ultimately you and I and everyone else may be the same being in many different bodies so there are seven billion of us here now but we are all one.

# Chapter 35

# The Greatest Opportunity

As individualisations of conscious Life that have incarnated in atomic and non-atomic bodies we have an incredible opportunity. We could be here for training to improve the operating system of the Universe as a whole and as we grow the Universe grows with us because everyone is connected always and everywhere in everything. But we can also choose to become entirely new, eternal, individualised beings in the super-physical level of the Universe. The beauty of this process is that each human being seeded in Earth is totally unique. It is our unique personality forged through the diversity of physical circumstances that makes us special.

Because of the amnesia we are in, not knowing who we really are where we come from or where we are going to, there is an innocence about us. We are, if you can excuse the relingo, baby gods gestating in a physical womb. We have an awesome opportunity that few realise.

By analogy, a ship is full of crew headed by a captain. However, every crew member has the opportunity to work and serve until they become a captain. To get a gist of this potential I would refer you to *The Keys of Enoch* by J.J. Hurtak.

To realise the enormity of ascending beyond heaven and hell into the super-physical realms of the Universe we have to understand the opportunity. Many people have different theories and complicated ideas but Yeshua made it very clear in a series of channelings on ascension that I disseminated in the early 1990's.

First we need to have a belief in the Source whatever we perceive the Source to be. I have done my best so far to help people who don't like religion to come to terms with the Source of our being as life. It shouldn't be difficult to believe in Life because we are alive!

## Chapter 35 – The Greatest Opportunity

The second requirement is we need to be open to the possibility of ascension. Because we have free will ascension can't happen to us unless we allow it. Obviously we would have to accept the possibility of continuous living because ascension is continuous living at a higher level than heaven.

The third and most important criterion for ascension is living out of love. We have to make every effort we can in our daily lives to practice unconditional love and forgiveness for all living things.

There is nothing elitist about ascension. It is the birthright of every human being. If you cast your mind back to the experiences of Storm and McCormack, they were rescued by a recognizable human in the ascended state with enormous power. Can we be trusted with that sort of power? Any negativity, selfishness, spitefulness, anger, resentfulness in us and we could reap havoc!

I believe there is no better situation for us to reveal our true colours and our hidden potential than where we are right now. The challenges of human life are for training and testing potential managers for the expanding Universe. The human condition is not about achieving heaven over hell but ascension beyond both polarities into super-physicality. We are here not just to improve the skills of the existing Universal managers but to become new managers ourselves!

However the super-physical cannot be taken by force or ambition. It cannot be stolen or seized. The secret is humility. The path to the super-physical is one of letting go and allowing the mystery to grow rather than gunning for it and trying to achieve and succeed as though we were going for a competitive career on the corporate ladder. The way is to let go of ego and ambition and surrender to the light.

The living light is conscious. It is alive. It is both powerful and gentle, subtle and supreme and the one who channels it must be sensitive to it and sensitive to Life as a whole. The managers don't dictate or boss; they serve. They serve the living light of Life and use it to serve the best interests of every other living thing they encounter. Life doesn't need the managers; the managers need Life. Attaining to the living light of Life is not an opportunity to conquer it is an opportunity to love and to serve.

# Continuous Living

In an electric circuit of a computer components aren't there for themselves, they are there to serve the function of the computer as a whole. In the Living Universe every component is alive. Every living component is a sentient being acting for the whole, not for itself.

Accepting a teacher is part of the process because it is the way we learn how to be humble and serve selflessly. We don't have to stick with a teacher. We can learn from one and move onto another, returning to the first at will as each teacher has a different personality but they all act as one.

A true teacher will teach us to let go to the light and play because more than anything serving in the super-physical is play, play with light; as one would play a musical instrument and play, as one would play as a child. The way is subtle, and the path is sweet and every step upon it is both the beginning and end to reach. People who are too proud to take on a spiritual teacher are like kids who think they can get a degree without a teacher. Some super clever kids can but most of us would fail dismally. If you can manage on your own fine but if you can't put out for a mentor.

Eventually we need to move beyond the need for a teacher altogether. Attachment to teachers can be a trap that will hold us back in our development. True teachers encourage devotees to move on. Just as a child needs to let go of its parents to become and adult so we need to know when to let go of a teacher and face the world alone and unaided. That takes courage.

A great impediment on the path is pride. A great opportunity is to learn humility. Pride is attachment to the mind. Humility is detachment from the mind. Pride is the way of the head. Humility is the way of the heart.

Some people find their world falls apart when they embark on a spiritual path. They lose their job. The wife runs off with the window cleaner. The dog bites a lawyer and the teenage son goes for a spin in the brand new Audi, uninsured of course, and crashes it. By the time the wreckage is sorted by a series of miracles, pride is replaced with humility!

# Chapter 35 – The Greatest Opportunity

The path finds the person. It is not easy to explain but if we are open to the way it will open up before us. We may meet someone or read something that will resonate with us.

We never have to force the way because force closes the way. If we respond to a nudge the way opens up to us. It will come and greet us. It may even seduce us. The secret is to recognise it when it comes and that we can only do with our hearts not our minds, with our feeling rather than our reasoning.

When the student is ready, the teacher will come and the teacher can come in many forms both human and animal and plants can be teachers too. Situations can teach us, as well as unforeseen circumstance. The most powerful teachers are the challenging ones, the difficult people and the problematic situations, the misadventures and misfortunes. If we can accept all of these as mentors and become masters of our destiny instead of victims of circumstance, that is our greatest opportunity because the way to mastery is self-mastery.

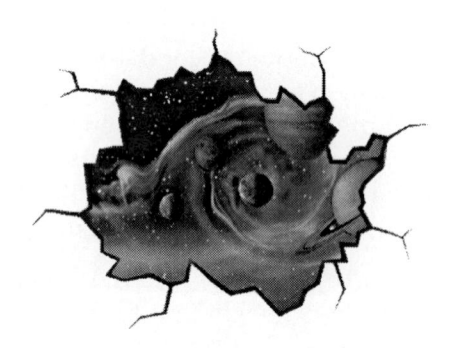

# Chapter 36

# My Mythology

Many people question the purpose of human life. I believe it could be to upgrade the operating system of the Universe by improving the existing management and also by creating new managers. In *The Intelligent Universe,* Sir Fred Hoyle, – who I admire as one of Britain's greatest cosmologists – said he didn't like the idea of God as creator of the Universe but he did like the Greek idea of the gods as managers in an already existing Universe. My sentiments entirely!

In Greek and Roman mythology the gods appear to be a fickle lot. Some would argue they behave like people because they are a projection of people's imagination. But equally it could be argued people are a projection of the imagination of the gods because my mythology is that human beings originate from ancient, non-atomic forms of Life – corresponding to the gods or the angels – that have projected their conscious personalities into bodies, physical and extra-physical, onto the Earth, in order to become fit for purpose as managers in the Universe. The idea was in the verse of a song I wrote in 1995:

*We are the gods, here in human form,*

*Here to overcome through battle, pain and storm,*

*That tenacious hold onto anger, greed and pride,*

*That prevents us as gods from evolving on high.*

My mythology is that the Earth is a flight simulator for pilots of consciousness. We enter the simulator for training by donning a physical and extra-physical body through which we can experience the simulated realities as though they were real. Each lifetime on Earth represents a session in the simulator. Another way I express this idea is to describe the Earth as a Womb of Angels. We are gestating angels. I got that idea from Emanuel Swedenborg who said that "Angels are not created in heaven by God but arise from the souls of deceased humans." According to

# Chapter 36 – My Mythology

Christian Morgenstern, the primary *modus operandi* for training burgeoning gods is through suffering and difficulty, challenge and adversity:

*"Your desire no more to suffer causes only new pain,*

*Thus will you never shed your garment of sorrow,*

*You will have to wear it until the last thread,*

*Complaining only that it is not more enduring,*

*Quite naked must you finally become,*

*Because by the power of your spirit,*

*Must your earthly substance be destroyed,*

*Then naked go forward in only light enclosed,*

*To new places and times, to fresh burdens of pain,*

*Until through myriad changes a god so strong emerges,*

*That to the sphere's music you your own creation sing.*

The physical world will be full of pain if we try to escape the lessons of humility and compassion, selflessness and courage that suffering can bring. But Life doesn't have to be painful. It can be full of joy if we let go of our resistance and learn through our own suffering to have compassion for others.

If we get involved in judgment of good and evil we may remain in heaven or hell forever. We will never be ready to ascend into the super-physical and birth as angels while we are caught up in the simulated issues of duality.

Unconditionality comes if we move beyond judgment to love everyone and everything, including the Ancient of Days. Wisdom comes of understanding and understanding the Ancient of Days comes from appreciating the meaning of his other appellation: *Gate Keeper of the Beyond the Beyond.*

The beyond is the extra-physical and the beyond the beyond is the super-physical. As a gate keeper for the super-physical he may not be intrinsically evil. He may be just doing his job as one charged to ensure that beings do not enter the super-physical levels of the Universe unless or until they have proved worthy to wield the power of the living light; a worthiness that comes from humility.

Everything in our world is a setup. Nothing is real because it is a virtual reality. You may scoff at me. Howard Storm used to dismiss the likes of me but not anymore. Since his NDE he says the physical world is like a dream. Stepping out in the extra-physical body is like an awakening. What we perceive to be evil may be a test to bring out the best or the worst in us.

If Satan guards the gates, to enter the super-physical we have to get past him. If we are in fear of him we will turn and run but if we are fearless and love everyone and everything that exists as a manifestation of Life then we will beam him a smile and he will bow in response and open the gate. All he is looking for is a loving and a guileless heart. His role has been to test and tempt us to see if we can really live and act out of love, despite whatever he and his cohorts throw at us!

## So who is Satan?

In Hinduism, *Sanat Kumara* is revered as the eldest son of Brahma, the creator. Sanat is also known as 'Skanda', son of Shiva. Portrayed as ever youthful he is considered to be one of the progenitors of mankind. According to a number of sources, including Mark Prophet, Sanat Kumara – kumara means prince – is known as the Ancient of Days. If that is true then through the commonality of this name we see a link between Sanat Kumara and Jehovah because the Ancient of Days described by the prophet Daniel in the Bible was the same god that appeared to Moses in the burning bush and declared himself as YHWH sounded by some as 'Yahweh', or 'Jahveh', commonly pronounced Jehovah.

Sanat Kumara is also known as the 'Planetary Logos', which means the 'God of this world.' Many – including the Cathars and Christians, Jews and Muslims – consider the 'God of this world' to be Satan. There is a link between Satan and Sanat. If you look at the two names side by side you will see one is an anagram of the other.

If you approach him with love, Sanat Kumara is a wise and loving being. If you approach him with fear or pride he will snap into his anagram and mirror that back at you a million fold.

# Chapter 36 – My Mythology

The second son of Brahma, called *Sananda Kumara*, is believed by many to be Archangel Michael. Jehovah Witnesses and Seventh Day Adventists believe that Jesus was an incarnation of Michael. The two 'fathers' Jesus referred to could have been Lucifer and Michael, Michael being the heavenly, super-physical father who seeded him.

A similar mythology occurs in the ancient Sumerian texts that predate the Bible. Abraham was a Sumerian. In the Sumerian legends two brothers, sons of *Anu*, called Enlil and Enki came to the Earth from space and bioengineered humanity as slaves to mine gold in Africa and labour in a garden between the Tigris and Euphrates called *Edin*.

In Hinduism Sanat Kumara is not considered to be evil. In the East they believe in polarity but not in absolute evil because everything is thought to work for the greater good. Lucifer, who is an aspect of Satan, means the light bearer.

In my mythology Sanat and Sananda, with their Kumara brothers Sanaka and Sanatana – corresponding to the Archangels, Gabriel and Raphael – operate the flight simulator. Working for the greater good of humanity I believe they hold the polarities of good and evil, light and dark, love and fear in the simulator for pilots of consciousness.

In the Hindu mythology, Brahma's sons refused to go out to spread life in the early Universe. Three times they were asked to go out to do what they were created for and three times they refused. On the third refusal, rage rose from the heart of Brahma and exploded out of his forehead as Shiva the destroyer.

Shiva drove the recalcitrant sons out – they all fell from grace – but Shiva continued to rage destroying Brahma's creation and so Vishnu the preserver came into being to maintain the balance between creation and destruction. There is a profound teaching in this mythology. In the Hindu religion the opposite polarities of creation and destruction are not viewed as good and evil but as polarities that need to be balanced. The Hindu trinity of the godhead represents this balance. In balance, neither good nor evil predominates, but between light and dark equilibrium occurs. In the West the wisdom of balanced polarities has yet to be fully appreciated.

# Continuous Living

In the Indian mythology the sons of Brahma had a challenging time. Their progress was fraught with difficulty and adversity but they established the templates of biological life and were all restored to the bosom of the father, including Sanat Kumara.

With the expansion of the Universe and the proliferation of Life, a problem arose. The individualisations of Life – call them gods or angels – could not compare with the original team in management skills. So the archangels were called to bring their progeny – the legions of angels – up to their own standard.

The growth in spirit of the original team of elohim came through the difficulty, challenge and adversity they had to overcome when they were banished and separated from their father. Separation was the situation they decided to simulate on planet Earth for us.

In my mythology biological life was established on the watery womb like planet Earth through a process of creative evolution, and genetic engineering, for the emergence of human kind. Physical bodies, with brains acting as filters to establish amnesia from other levels of reality, were set up to simulate the illusion of separation from source.

Lured or coerced into physical bodies, subject to toil and hardship, pain and disease, parasites, predation and death, gave baby angels reasons to cry. Life in the jungle with snakes and insects and poisonous plants and spiders, not to mention voracious animals was a good start to the angelic descent into adversity. The challenge was set. The rest is history.

But why the Earth; why out of trillions of planets in billions of stars in our galaxy alone was the Earth chosen as a womb of angels, a boot camp for gestating gods? I believe it is because of the $23.5^0$ list of the Earth on its axis. I believe the accidental tilt of the Earth toward the sun set up a means whereby the planet would be periodically purified.

Periodic catastrophes set terms at the planetary school for gods. At the end of each age the planet could be cleared of populations and civilisations, ready for a fresh start.

# Chapter 37

# The End of an Age

It is important that we don't take our lives for granted. Great changes could be upon us at any time. Throughout my life I have had a deep sense we are in the *End Times,* that we are fast approaching the end of an Age. This premonition was reinforced when I first read an extraordinary book called *Worlds in Collision*[1] by Immanuel Velikovsky.

I was a student and the year was 1969. When I reread his book on catastrophes in 2014, it struck me as highly unlikely that the historical events Velikovsky reported could have been caused by massive 'worlds in collision' because humanity wouldn't survive disasters of such a magnitude. Even a large asteroid or comet colliding with the Earth would cause devastation such that few would live to tell the tale and yet Velikovsky's historical research told of populations in ancient times surviving a repeating pattern of world disasters associated with **a reversal of the direction of the sun in the sky.**

Velikovsky delved deep into the ancient records and discovered in Herodotus' second book of history, reference to a conversation with Egyptian priests in which they asserted that since Egypt became a kingdom,

> *"...four times in this period the sun rose contrary to his wont; twice he rose where he now sets, and twice he set where he now rises."*[2]

Pomponius Mela, a Latin author of the first century wrote, *"The Egyptians pride themselves on being the most ancient people in the world. In their authentic annals one may read that the course of the*

---

1 **Velikovsky I.,** *Worlds in Collision,* Victor Gollancz Ltd., 1950
2 **Herodotus,** Bk ii 142 (transl. A.D. Godley 1921)

*stars has changed direction four times and the sun has set twice in that part of the sky where it rises today."*[3]

The Papyrus Harris speaks of a cosmic upheaval when..."*the south becomes north and the Earth turns over.*"[4] The Papyrus Ipuwer stated *"The land turned round as does a potter's wheel and the Earth turned upside down."*[5] In the Ermitage papyrus reference is made to a catastrophe that turned the Earth upside down.[6] The sun was known in ancient Egypt as 'Harakhte'. There can be no doubt that the catastrophe was associated with a reversal in the direction of the sun because elsewhere in the Ermitage papyrus there is reference to the original direction of sunrise *"Harakhte he riseth in the West."*[7] Texts found in the pyramids say *"the luminary ceased to live in the occident and shines anew in the orient"*.[8]

Velikovsky went on to claim in the tomb of Senmut, architect of Queen Hatshepsut, a panel on the ceiling shows the celestial sphere with the signs of the zodiac and other constellations in a reversed orientation of the southern sky where north is exchanged for south and east for west.[9]

The simplest conclusion to draw from this anomaly is that this was how the night sky appeared to the architect in his day? Velikovsky cited Plato in 'Politicus', *"I mean the change in the rising and setting of the sun and the other heavenly bodies, how in those times they used to set in the quarter where they now rise, and used to*

3   **Mela** Pomponius, *De Situ Orbis.* i. 9. 8.
4   **Lange** H., *Der Magische Papyrus Harris* Danske Videnskabernes Selskab, (p58) 1927
5   **Lange** H., *German translation of Papyrus Ipuwer 2:8,* (pp 601-610) Sitzungsberichte d. Preuss. Akad. Der Wissenschaften 1903
6   **Gardiner,** *Journal of Egyptian Archeology I,* (1914); Cambridge Ancient History I, 346
7   **Breasted,** Ancient Records of Egypt III, Sec 18.
8   **Speelers** L., Les Texts des Pyramides I, 1923
9   **Pogo** A., The Astronomical Ceiling Decoration in the Tomb of Senmut, Isis (p.306) 1930

## Chapter 37 – The End of an Age

*rise where they now set."*[10] The reversal of the sun in the sky was never a peaceful event, Plato continued in 'Politicus', *"There is at that time great destruction of animals in general and only a small part of the human race survives."*

Velikovsky found in the drama *Thyestes* by Seneca a powerful description of what happened when the sun turned backward in the morning sky[11], which Plato detailed in *Timaeus* (probably from Pythagoras) *...a tempest of winds... alien fire...immense flood which foamed in and streamed out* (tsunamis)*...the terrestrial globe engages is all motions, forwards and backwards and again to right and to left and upwards and downwards, wandering every way in all the six directions.*[12]

In the sacred Hindu book 'Bhagavata Purana' four ages or 'Yugas' are described, each ending in a cataclysm in which mankind is almost destroyed by fire and flood, earthquake and storm.[13]

If these age changing events were in the historical record of a single culture we could dismiss them but they are documented by almost every culture in the world from a time when they were totally disconnected. Hesiod, one of the earliest Greek authors, wrote about four ages and four generations of men destroyed by the wrath of the planetary gods[14]. He described the end of an age as *"The life giving Earth crashed around in burning...all the land seething, and the oceans...it seemed as if Earth and wide heaven above came together; for such a mighty crash would have arisen if Earth were being hurled to ruin, and heaven from on high were hurling her down."*[15]

---

10   **Plato,** *Politicus* (transl. H.N.Fowler, pp 49 and 53, 1925)
11   **Seneca,** *Thyestes II,* ( transl. F.J.Miller 794 ff)
12   **Plato,** *Timaeus,* (transl. Bury, 1929)
13   **Moor** E., The Hindu Pantheon 1810
14   **Hesiod,** *Works and Days* (transl. H. Evelyn-White 1914)
15   **Hesiod,** *Theogony* (transl. H. Evelyn-White 1914)

# Continuous Living

The Persian prophet Zarathustra spoke of, *"...signs, wonders and perplexity which are manifest on the Earth at the end of each age."*[16]

The Chinese call the perish of ages 'Kis' and number ten Kis from the beginning of their known world until Confucius.[17] In the ancient Chinese encyclopedia, 'Sing-li-ta-tsiuen-chou' the general convulsions of nature are discussed. Because of the periodicity of these convulsions the Chinese regard the span of time between two catastrophes as a 'great year'. As during a year, so during a world age, the cosmic mechanism winds itself up and...*in a general convulsion of nature, the sea is carried out of its bed, mountains spring out of the ground, rivers change their course, human beings and everything is ruined, and the ancient traces effaced."*[18]

An ancient tradition of world ages ending in catastrophe is persistent in the Americas amongst the Incas[19], the Aztecs and the Maya[20]. The Maya have a 'Long Count' between global catastrophes similar to the Chinese 'Great Year'.

In her book on global catastrophes in ancient times,[21] Lucy Wyatt suggested a date for Noah's Deluge seventy nine years after the beginning of the Long Count that ended in 2012. Her research indicated that cataclysms occur close to the target date of a Long Count or a Great Year rather than on it. That would fit with the principle in science of 'margin of errors'. In a cycle of global cataclysms millennia apart one would expect a cataclysmic event to occur sometime within a margin of a few years of a predicted date.

The Maya predicted each age ended with earthquakes at the solstice. It is the Maya reference to global earthquakes and the

16 **Muller** F., ed. Pahlavi Texts: The Sacred Books of the East, 1880
17 **Murray** et al, An historical and Descriptive Account of China, 1836
18 **Murray** et al, An historical and Descriptive Account of China, 1836
19 **Alexander** H., Latin American Mythology, 1920
20 **Humbolt** A von, *ResearchesII,15*
21 **Wyatt** L., *Approaching Chaos*, O Books 2009

## Chapter 37 – The End of an Age

solstice that provides the clue to understanding how predictable pole shifts could occur without assuming the planet goes *head over heels* in a world collision that would be totally unpredictable.

Historical records from every continent report that the world has fallen over the poles at least four times in the memory of mankind. A major part of stone inscriptions found in the Yucatan refer to world catastrophes. The most ancient of these *katun* calendar stones of Yucatan refer to great catastrophes, at repeated intervals, convulsing the American continent. The indigenous nations of the Americas have preserved a more or less distinct memory.[22] In the chronicles of the Mexican kingdom it is said, *The ancients knew that before the present sky and earth were formed, man was already created and life had manifested itself four times.*[23]

The sacred Hindu books, the 'Ezour Vedam' and the 'Bhaga Vedam' share the scheme of expired ages, the fourth being the present. They differ only in the time ascribed to each age.[24] The Buddhist 'Visuddhi-Magga' describes seven ages, separated by world catastrophes.[25]

A tradition of successive creations and catastrophes is found in Hawaii[26] On the islands of Polynesia there were nine ages recorded and in each age a different sky was above the Earth.[27] Icelanders believed that nine worlds went down in a succession of ages, a tradition contained in the 'Edda'[28]

22 **Brasseur de Bourbourg** C., S'il existe des Sources de l'histoire primitive du Mexique dans les monuments égyptiens, 1864
23 **Brasseur de Bourbourg** C., Histoire des nations civilisées du Mexique I, 53, 1857-1859
24 **Volney** C., New Researches on Ancient History (p.56) 1856
25 **Warren** H., Buddhism in Translations 1896
26 **Dixon** R., Oceanic Mythology, 1916
27 **Williamson** R., Religious and Cosmic Beliefs of Polynesia I, 89, 1933
28 **Völuspa,** *The Poetic Edda,* (transl. H. Bellows 1923)

# Continuous Living

There are seven ages in the rabbinical tradition of 'creation': Already before the birth of our earth, worlds had been shaped and brought into existence, only to be destroyed in time. This Earth too was not created in the beginning to satisfy the divine plan. It underwent reshaping, six consecutive remouldings.

New conditions were created after each of the catastrophes...we belong to the seventh age.[29] According to the rabbinical authority Rashi, Hebrew tradition knew of periodic collapse of the firmament, one of which occurred in the days of the Deluge.[30] The Jewish philosopher Philo wrote, *"Great catastrophes changed the face of the Earth. Some perished by deluge, others by conflagration."*[31]

In Isaiah 24:1 is written, *"Behold the Lord maketh the earth empty, and maketh it waste, and turneth it upside down, and scattereth abroad the inhabitants thereof"*.

Muslims throughout the world are taught that the Day of Judgment is imminent and when it occurs the sun will change its direction in the sky. Muslims believe a day of divine retribution is coming when the sun will set in the East instead of the West and will rise in the West instead of the East. They have been told that they will know this time when the wandering and nomadic people (the Arabs) will leave their wandering and nomadic way of life to settle in exceedingly high buildings (Dubai) and when the slave gives birth to her mistress – a time when daughters will treat their mothers as slaves!(sounds familiar?)

Reports of the Earth turning upside down over the poles are difficult to take seriously but coming from so many and diverse sources they are also difficult to dismiss. As I developed the cosmology of the vortex I realised something could be happening to the Earth that would make sense of these otherwise bizarre historical records.

---

29   **Ginzberg** L., *Legends of the Jews*, 1925
30   **Rashi,** Commentry to Genesis 11:1
31   **Philo,** *Moses* II, x, 53

# Chapter 37 – The End of an Age

Two things are happening to our planet. One is global warming. The other is a weakening of the magnetic field[32] and I don't believe they are coincidental. With the Earth having an iron core I think they are linked because it is common knowledge that heating iron weakens magnetism.

Greenhouse gas emissions are thought to be the cause of global warming but that theory cannot explain the global warming of Neptune[33]. An altogether different account for global warming comes out of the vortex theory that would cause the increase in global temperature to originate from the core rather that the surface of the planet.

An increase in the temperature of the iron core of the Earth could weaken the Earth's magnetic field and when the Earth's magnetic field weakens a magnetic pole reversal would be most likely to occur[32]. A magnetic pole reversal during a period of global warming could also make sense of Velikovsky's descriptions.

The interior of the Earth is hot. The heat is caused by the immense pressure due to gravity. According to my vortex theory energy released by gravity is linked to the annihilation of matter and antimatter through zero space; a point of singularity which, at the gravitational centre of the Earth, could be acting as a mini black hole.

After annihilation the energy released would be captured in it, in vortex motion, until it overcomes the gravity. Then it could leak out. Building up energy until it escapes then repeating the process is how a geyser works. In *The Vortex Theory* I use the gravitational geyser model to account for periodic gamma ray bursts from galactic cores. A geyser action of planetary core gravitational energy could account for the periodicity of global

---

32   www.extremetech.com
33   **Hammel**, H., and Lockwood, G., Suggestive correlations between the brightness of Neptune, and Earth's temperature, *Geophysical Research Letters,*

warming and cooling, weakening of the global magnetic field and pole reversals. But why would it cause pole reversals?

We live on the crust of the Earth which sits on an outer mantle of molten rock that secures it to the inner planet. An increase in temperature of the molten mantle due to energy escaping from the core would decrease its viscosity. One could imagine it somewhat like honey that turns from sticky to runny when it is warmed. The crust of the Earth is effectively glued in place by the stickiness of the outer mantle. Were this to warm up and become lubricant the crust might then float. Then it could slip from its present position.

If the crust of the Earth were to slip over the outer mantle that could cause the events described by the ancients without the entire globe going for a tumble as imagined by Velikovsky because the poles would reverse under the crust. If the pole positions on the crust reversed, by slipping over the actual poles of the turning globe, the east-west orientation on the crust would also reverse.

A slide of the Earths crust over the poles of the planet would explain the claims from the ancient past that the sun periodically changes its direction in the sky. However, a slip of the Earth's crust would result in shock waves that would cause unprecedented earthquakes and tsunamis throughout the world.

There would also be extreme heat, conflagrations and violent storms in the atmosphere with high ocean evaporation followed by deluge, due to the heat generated in the Earth's crust by friction as it moves. That could account for the worldwide catastrophes recorded by the ancients, which accompanied the purported reversal in the direction of the sun in the sky.

But why would a floating crust slip over the poles? The answer lies in the 23.5º list of the Earth responsible for the seasons. At the summer and winter solstice our planet is titled toward the Sun. At that time one pole is closer to the Sun than the other. The difference in pull of the gravity of the Sun, on the opposite north and south caps of a floating crust could cause the crust to slide over the poles of the planet.

## Chapter 37 – The End of an Age

If the crust were to slip over the poles, to people on the Earth it would appear that the poles have reversed. If this crust slip hypothesis were true, it would be highly unlikely that the crust would resettle in the same position over the poles of the planet. That could explain why Antarctica was once in the tropics and Hudson Bay was once over the North Pole.

We are in the midst of a phase of global warming so the vortex theory predicts that we are approaching another crust slip and consequent pole reversal. In 2014 the European Space agency predicted an imminent magnetic pole reversal but by imminent they meant any time in a few thousand years. However, an article in *Scientific American* suggested a reversal of the poles could happen much sooner than expected.[34]

The crust slip hypothesis links a number of seemingly disconnected events that occur periodically and the geyser model for gravitational energy release from black holes would account for the periodicity. Periodic ice ages interspersed with global warming and periodic magnetic pole reversals are in the scientific annals.

Periodic global catastrophes accompanied by periodic reversals of the direction of sunrises and sunsets are in the historical record. All could be a consequence of periodic releases of gravitation energy from the core of the planet. I like to describe it as the *heart beat of the Earth*. This presupposes the gravity we experience is due to a vortex interaction between matter and antimatter through the centre of the Earth.

Annihilation energy rising periodically from the centre of the Earth would explain the Hindu tradition of Shiva the destroyer awakening to end a Yuga and herald the dawn of a new age. So many great civilisations have vanished without trace. Maybe that is because none could survive a beat of the planetary heart.

---

34  http://www.scientificamerican.com/article/earth-s-magnetic-field-flip-could-happen-sooner-than-expected/

# Chapter 38

# The New Age

In 1985, during his NDE, after he had been rescued into the Light, Howard Storm was told by his teachers that the Cold War would end. He was skeptical and replied "Maybe in a couple of hundred years." The response he got was no, it would be more like a couple of years. Then he was told that the world was at the beginning of a major transformation, a spiritual revolution that would affect every person in the world. When he asked what it would be like he was transported into the future. He was given a vision of a beautiful natural wooded setting.

He saw no evidence of human intrusion or manmade devices. They told him this was the future. He was in a garden that people tended. People coming by and talking with each other were dressed simply but with exotic ornaments. They resembled Native Americans in their dress. Storm enquired about what they did. He was told they raised children mostly. Most people spent the majority of their time with the children teaching them about nature and the wonders of the natural world.

Howard Storm was told in the future people would not make a distinction between work and play but would all participate in rearing and teaching children as preeminent activity. He was shown people raising food by sitting next to plants and communing with them. In minutes they were able to harvest mature fruit and vegetables. They ate what they grew immediately without cooking. The clothing was finely woven natural fibers. There was very little metal except in their ornaments.

Howard asked if this was paradise. His teachers laughed, "No, only compared to the world you live in." He was told that in this future world people would still have sickness but the treatment for disease would normally be successful. People would gather round the person in need of help and through prayer, touch and

## Chapter 38 – The New Age

meditation the disease will be cured. He was told in the future people will grow only enough food for their needs,.

Collectively all the people of the world will control the weather. The climate will be regulated by the collective will of humankind. The plants will be loved and raised by individuals. All animals will live in harmony with people. There will be countless small communities all over the world and each will have its own identity and culture.

There will be many different languages but people will be able to communicate telepathically. There will be no technology because there will be no need for devices since humans will have the power to control matter and energy. Storm was told people would stay within their communities unless they wanted to experience life in another culture with different music, or vegetation or scientific investigation.

Howard Storm was amazed because the future was so different to the images he anticipated from science fiction. Instead he saw people living in extreme simplicity, like indigenous peoples. There was harmony, with no war or want and everyone seemed to be happy.

Howard Storm was told this is what the world would be like in a couple of hundred years. He expected the Berlin wall to come down then, not the total eradication of global civilisation!

We all know that the civilisation sprawling over the Earth, with the world population rising, is unsustainable. It cannot last forever. Some of our greatest Scientists, including Astronomer Royal, Lord Rees, say the 21st is our final century.[1] For this prediction Rees received the Templeton Prize.

In terms of recovery it would be better for the Earth and the future of mankind if civilisation and unsustainable levels of population disappeared sooner rather than later. In the Montauk

---

1    **Rees** Martin, *Our Final Century,* Arrow 2004

# Continuous Living

Project[2], Preston Nichols predicted that in the 21st Century the world population would be reduced from seven billion to three hundred and fifty million or thereabouts and civilisation would mostly be gone.

The Eastern Indians speak of the cycles of four ages or Yugas running from Sat Yuga – the first age of greatest light – into Kali Yuga – the fourth age of greatest darkness. We are supposedly in a Kali Yuga so the next great age should be a Sat Yuga. That fits with what Howard Storm was shown. It makes sense that a period of planetary recovery should follow a Kali Yuga.

In my previous title, *Activation for Ascension*, I spoke of a 26,000 year cycle split into two 13,000 periods. In the book *When The Earth Nearly Died*[3] this time span was put closer to 11,500 years. The Adam and Eve story occurred in my opinion at the beginning of the most recent 13,000 to 11,500 year period when the decline away from Sat Yuga toward Kali Yuga began. That was described in Western religions as the *fall from grace*. I believe the deluge of Noah occurred somewhere between five and seven thousand years ago. This model fits with the emphasis on four ages in the ancient records with mention of cycles of ages before that.

The succession of ages from light to dark and then back to light again fits with the Yin Yang principle that when something comes to its completion it changes to its opposite.

It is an ancient tradition that the cycle of ages is more a spiral than a circle. With each turn of four ages from Sat Yuga to Kali Yuga, humanity moves to a higher level of evolution. This is part of a greater Universal cycle wherein the Universal mind in growing in wisdom.

The quantum or atomic principle is that the whole is made of parts. The collective mind of the Universe could be gaining

2    **Nichols** Preston, *The Montauk Project* Sky Books 1991
3    **Allen** D *When the Earth Nearly Died* Gateway Books 1995

# Chapter 38 – The New Age

through the collective experience of all expressions of Life throughout the Universe. That would include, in no small measure, the human experience on Earth.

I summarised my perception of the plan of ages in a song I wrote for my sister Jenny, in 1996, when she was going through a particularly difficult time in her life:

## The Plan of Ages

*Somewhere hidden deep in every human heart,*
*Lies the spring of Life, the eternal spark,*
*The door to heaven stands open wide,*
*The door is in the heart so step inside,*

*Step within the breath you breathe,*
*Breathe in deep now and feel the peace,*
*Let the breath lead you into your heart,*
*Let it connect you to your eternal spark.*

*The plan of ages slowly unfolds,*
*Within its great purpose man evolves,*
*Ages of suffering grim to bear,*
*But how else would we learn to love, care & share,*

*So let not your troubles hold you in their sway,*
*Nor let your problems sweep you away,*
*These are but the trials that make us strong,*
*They'll keep on coming while to Earth we belong.*

*So breathe your way through every trial and task*
*Knowing that none will ever last,*
*And breathe into every good fortune and joy,*
*Knowing also that they will pass,*

*Our lessons they come in light and dark,*
*Let neither trick you from your eternal spark,*
*Good and evil are but the masks,*
*That life wears in the play of human hearts.*

# Chapter 39

# Conclusion

In every generation both polarities are represented by incarnations on Earth. In recent centuries the polarity of the dark has become more vociferous due to the advance of scientific materialism. This false philosophy has been at the core of capitalism and communism. Materialism is fed by greed and driven by the power of money.

In earlier centuries and even up to our day, the false philosophy of monotheism has operated as another agent of the polarity of darkness, evident in the cruelty and carnage it always leaves in its wake. The rise of the Islamist cult in the name of the monotheism of Abraham is evidence of this.

Reason can maintain philosophies both true and false. Reason is used in both monotheism and materialism to recruit followers and to justify inhumanity. Falsehood is revealed if the reasoning is based on a pack of lies. As well as lies, the hallmark of monotheism and materialism is war.

The followers of these false philosophies will use reason to warp morality and justify murder. In France, for example, the false morality of monotheism justified the crusade against the Cathars and immorality in materialism justified the reign of terror after the French revolution. People who believe they are right will stop at nothing to maintain their self-righteousness and use every means possible, even murder, to maintain their limited point of view. Such is pride!

My friend Vincent said the root problem in history has been the use of human reason to bring things into existence rather than allowing Life to bring things into existence. Life allows things to unfold. It is a flow that enables things to happen naturally.

Reasoning leads to control and domination, laws and bureaucracy, rules and regulations whereas joyful spontaneity allows for magic. Reason resonates with the head and Life

# Chapter 39 – Conclusion

pulsates with the heart because reason is in the domain of thinking whereas Life is more about feeling and simply being alive.

If an idea pops into your head or comes from another source note how you feel when you first entertain it. Let go of the thought if it feels bad. That can be hard because negative thoughts tend to be like sticky grass.

We have an initial subtle feeling that is fleeting then reason barges in with some sensibility that contradicts the intuition. The result is usually disaster as the warning or promise of the heart goes unheeded.

I have a saying, good invites; evil invades. Notice how an inviting intuition from the heart is overridden immediately by an invading thought from the head.

What I like about Howard Storm's vision of the future is it seems a lot more heart than head, more in tune with Life than the rule of law. If the future on Earth turns the way he saw in his vision, then humankind may be about to undergo a transition away from science and technology, government and commerce, industry and electricity, cities, roads and railways into a much simpler way of life where people are more free to be spontaneously happy, living on the land close to nature, as their hearts desire, in a more caring society. Then indeed we will have Heaven on Earth!

# Afterword

Just over three thousand years ago, a young man in Egypt headed into the solitude of the desert on a vision quest. He fasted and prayed to know the structure of the Universe and the origin and destiny of mankind. Toward the end of his vigil, Poemandres – meaning the Shepherd of mankind – came to him.

Also known as the Logos or the Word, this Great Spirit appeared in his meditation and granted him his boon. He was taught that all worlds are formed of movement organised into distinct levels or planes. The name Hu-man was given to him for mankind meaning divine and mortal combined. He was told that in order to form humanity spirit descended down seven planes, into matter; the fall being not a punishment but an act of supreme love. Poemandres revealed to the young man that the Divine is Life and Light and that was the origin and the destiny of human kind.

Inspired by his vision, the young man returned from the desert and set out as a teacher and healer. In Egypt and across the Mediterranean in the emerging Mycenaean civilisation of Greece he set the scene for language, science, and medicine. After the decline of Egypt and Mycenae he became renowned more as a god of fluidity, Hermes or Mercury, than the man teaching that the state of fluidity – energy – underlies everything.

Nonetheless his legacy took root and his teachings were passed down into Classical Greece influencing philosophers from Pythagoras to Plato. In line with Hermetic teaching, Plato taught that the psyche – the conscious, thinking personality – was an immortal template that forged the flesh of man then departed into another domain when the mortal frame expired.

Aristotle, his most famous disciple turned against Plato by following the teaching of Democritus – whom Plato deplored. Aristotle embraced the teaching of Democritus that the psyche had no independent existence but was a mere by-product of the material body and perished with it.

# Afterword

The teaching of Democritus and Aristotle became the cornerstone of science. The Platonic idea of immortality of the psyche was never taken seriously in scientific institutions. Now, however, the idea that the individualisation of life, mind and consciousness might pre-exist and survive the physical body, and even act as a template for it, should be taken seriously because of advances in medical science.

Before World War II, Professor Harold Saxton Burr of Yale University Medical School published over thirty scientific papers on his ground-breaking research. Using high resistance voltmeters he discovered electromagnetic fields associated with living organisms that appeared to be responsible for their morphology[1]. Burr's idea of an electromagnetic template for Life aligned with Plato's concept of the psyche acting as a template for the physical body.

Burr's findings were supported by the Russian scientist Semyon Kirlian who, in 1939, discovered patterns in coronal discharges generated by high voltages that supported the idea of a field associated with living organisms that responded to electricity. Despite the fact he substantiated Plato's hypothesis according to the scientific method, Burr's research has been ignored because it undermines the materialistic world view of Democritus which is the quasi-religious belief of most scientists.

In 1975 Raymond Moody published Life after Life[2], documenting evidence of the continuity of human consciousness after death, arising from medical advances in resuscitation. Unfortunately his use of the phrase near-death-experience played into the hands of devotees of Democritus.

Sceptics, notably the psychologist Susan Blackmore, latched onto the word 'near' and attempted to explain away the

---

1     **Burr H**. Blueprint for Immortality, Neville Spearman1972
2     **Moody R**. Life after Life, Bantam Press 1975

experiences as the hallucinations of a dying brain[3]. Her arguments have been refuted by medical doctors at the front line of critical care and resuscitation medicine.

Dr Sam Parnia specialises in the resuscitation of people after they have died[4]. He insists the term near-death is inaccurate because teams of doctors and nurses in leading hospitals throughout the world are now using innovative techniques to keep the cells in a body viable after cardiac arrest when brain function has ceased and organs have failed.

When resuscitated these people come back, quite literally, from the dead. Parnia and other critical care doctors, routinely interview patients asking if they had any after or actual death experience (ADE). According to Parnia, throughout the world, people who have been documented with an ADE now number in the millions.

Experiences vary in detail but they mostly point to a similar conclusion. The conscious, thinking, feeling part of us can survive death to witness the scene of death before going off into another realm of reality. After death evidence is the strongest confirmation yet for continuous living.

Dr Parnia claims that science is dividing into two camps between those who claim only things we perceive through experiment and scientific observation of the known world pertain to science and those who are willing to listen to the witness of people who have been through a profound and often life changing experience, thanks to advances in medical science. But quantum mechanics is based on belief in unseen worlds full of virtual particles. It is the pre-eminent theory in science yet the uncertainty principle that allows for it is invalidated by experimental evidence pertaining to a real particle called the

---

3   **Blackmore S**. Consciousness, Hodder & Stoughton 2003
4   **Parnia S.** (with Josh Young) Erasing Death, Harper One, 2013

neutron[5]. Scientists, on one hand, explain away reports of people who have been resuscitated after death on the grounds the experiences pertain to unseen worlds then on the other hand they explain away evidence of real particles to protect their own belief in virtual particles in the unseen quantum world.

There is a rule in science that a hypothesis is believable if it can predict the outcome of a future observation. A core prediction in the Western religions is that a day would come when people would rise from the dead; that they would return from the metaphoric grave to tell us about heaven and hell. That belief in the resurrection of the dead has been around for thousands of years and now it is vindicated. Thanks to critical care resuscitation the resurrection of the dead is happening right now! The successful fulfilment of this prediction in religion also means to rubbish religion in the name of science is to demean science. Upholding the Caduceus, the symbol of Hermes, medicine has substantiated the message of continuous living given to him by Poemandres, on his vision quest three thousand years ago. The Word that became flesh a thousand years later, the Good Shepherd who proclaimed continuous living and resurrected Lazarus, was that the same Logos? Was that the Poemandres? Was he the selfsame Shepherd of Mankind who came in the vision to Hermes?

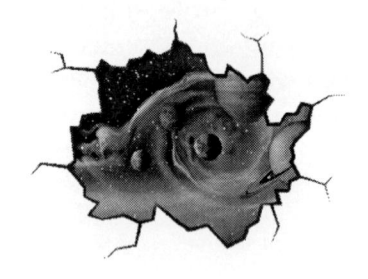

---

5     **Ash, D.** *The Vortex Theory* (Kima Global 2015)

# Appendix 1

# The Wisdom of Hermes

"*If then you do not make yourself equal to the Divine, you cannot apprehend The Divine; for like is known by like. Leap clear of all that is corporeal, and make yourself grow to a like expanse with that greatness which is beyond all measure; rise up above all time and become eternal; then you will apprehend The Divine. Think that for you too nothing is impossible; deem that you too are immortal, and that you are able to grasp all things in your thought, to know every craft and every science; find yourself home in the haunts of every living creature... but if you shut up your soul in your body, and abase yourself, and say, 'I know nothing, I can do nothing, I am afraid of earth and sea, I cannot mount to heaven; I do not know what I was, nor what I shall be'; then what have you to do with The Divine? Your thought can grasp nothing beautiful and good, if you cleave to the body and are evil. For it is the height of evil not to know The Divine; but to be capable of knowing The Divine, and to wish and hope you know The Divine, is the road that leads straight to the good; and that is the easy road to travel...for there is nothing that is not The Divine. And you say 'The Divine is invisible?' Speak not so. Who is more manifest than The Divine? Oh people of the Earth, men born and made of the elements but with the spirit of the divine man within you, rise from your sleep of ignorance! Be sober and thoughtful. Realize that your home is not in the Earth but in the light. Why have you delivered yourselves over unto death, having power to partake of Continuous Living? Repent; change your minds. Depart from the dark night and forsake corruption forever. Prepare yourselves to climb through the seven rings to blend your souls with the eternal light...when the ears of the students are ready to hear, then cometh the lips to fill them with wisdom.*" Corpus Hermeticum; XI.2

# Appendix 2

# Mind the Matrix

Free will to choose between the polarity of good and evil on planet Earth is a most precious gift. It allows for maximum learning and growth through choice in a matrix of opposites. I speak on this as *duality* in the film *Mind the Matrix* (on Vimeo.com).

*Mind the Matrix* highlights deceptions in civilisation and the machinations of those in power to exploit us. It then goes on to depict alternative lifestyles based on community living and individual ingenuity and creativity which point to the future that Howard Storm was shown where simple, rural communities would replace civilisation in the future.

It is easy to become obsessed with the evils of religion but if you were hunting humans like the evil God wouldn't you infiltrate the safe enclosure of religion and rip it apart with evil and atrocity to make your human prey more vulnerable to capture. The evils of corporations, banks and the military can also be bait to stimulate us to adopt a negative mindset to draw us away from unconditional love.

The world is meant to be both good and evil. If we look for the good in everything we will be drawn to the polarity of goodness. If we focus on the evil in everything we will be drawn to the polarity of evil.

*Mind the Matrix* reminds us of the dangers inherent in civilisation and the possibility of opting for alternative lifestyles which is good. Ultimately it is our ability to live out of love in both situations that really matters.

# Appendix 3

# Rebecca's Approach to Life

"*Life just keeps on going, free of feelings of suffering, joy, sadness, pity; like a river it tumbles each day into the next. How you wake up and perceive the day is up to you. How you choose to perceive others and yourself is also up to you. But notice how judgment and anger leaves you feeling inside; notice how when you don't communicate with people you are left feeling suffocated, blocked - even sick.*

*"Is the feeling of being right or better than others worth your own suffering? Notice then how when you give with love and understanding you are left with the feeling of joy and wholeness and there's no right or wrong, no good or evil, only a fresh new day with every sunrise. Finish each day knowing that everything is complete; there is nothing more to do or worry about. Relax into sleep in the knowing that there is nothing more important than the day that will awake you.*"

My daughter Rebecca sent me these words of encouragements in an email in March 2004 when I was going through a difficult time.

Rebecca has expressed concern about talk of an afterlife if it leads us to perform acts of care and kindness in order to get to heaven and avoid hell. She pointed out people should act with goodness from their hearts rather than from the selfish motive of attaining personal salvation. Rather than being concerned about an afterlife, she feels we should live for the life we have now. The answer to these concerns lie, I believe, in Rebecca's own words, "…when you give with love and understanding you are left with the feeling of joy and wholeness and (then) there's no right or wrong, no good or evil…."

# Appendix 4

# The Medicine Wheel of Life

Shamanism represents the wisdom teachings of the native peoples of the Earth. Native people have always honoured and respected Life and the life fields associated with living organisms. They have their own ancient traditions and ceremonies for establishing their connections with Life and the super-physical beings of the Universe. One particularly relevant tradition is the way of honoring Life in the seven sacred directions. In the Native American tribes this is called the *Medicine Wheel*.

In the Medicine Wheel ceremony the shaman goes to each direction in turn; the north, the east, the south and the west to call on Life, to pray and to offer gratitude. Those who are departing are remembered in the west. Those who are arriving are remembered in the east. Families and friends are remembered in the south. Ancestors are remembered in the north. Love and peace are sent to the nations in each direction. Grandmother Earth is honoured below and Grandfather Sky above. Finally, thanks are given to the spring of Life in the heart and in the breath.

In the community where I live, Anne leads the ceremony of the seven directions every day at noon in a wooded glade. As well as the standard ceremony she remembers the four primary archangels in each direction and honours the goddess of love, the goddess wisdom, the goddess of peace and healing and the goddess of power in each direction.

Everyone who performs the ceremony has their variant. What matters is that they do it, not how they do it!

# Appendix 5

# A Positive Approach to Dying

Some live in fear of departing this world and others make a dreadful fuss about it but if we believe in continuous living the transition we call dying can be treated as an adventure, the start of a journey, or a homecoming no different to taking off a heavy coat when we come out of the cold ready to step across a threshold into the welcome of a warm house. In the words of Banksy, *"When the time comes to leave, just walk away quietly and don't make any fuss"*[1]

My mother recently passed out of her physical body. She was looking forward to this time. She did not make a fuss about it. It was a labour for her, painful like the labour of bringing us into this world, but the outcome is joyful for her as she will be on the other side of a veil where she can help us all. She had a wonderful life on Earth and so it will continue for her as she sets off on her next adventure in continuous living.

Dying to this world can be a gift of Life because suffering terminal illness may be what it takes to bring about a change of heart through the sharp focus that only notice of termination of the physical body can bring.

Departing this world is the time when many people choose the direction they will take after shedding the mantle of mortality. The time of dying could be a critical time in physical life if a prime purpose of physical life is to choose the polarity of continuous living. If the choice is continuous living in the light, dying would be a time to cherish not to dread.

In traditional Chinese medicine, the soul has two parts, the Hun and the Po. The Chinese say that spirit is the in the Po and the emotions are in the Hun.

1    **Banksy,** *Wall and Piece,* Century, 2006

# Appendix 5 – A Positive Approach to Dying

If the person dying is not prepared for the transition, their emotions might remain attached to the physical plane. Instead of passing into higher realms with the Po, the Chinese believe the person can become Earth bound by their Hun, lost between the worlds, not knowing they have shed the physical body.

This is how the Chinese account for ghosts and haunting. The idea that people who have passed over can become earth bound, as ghosts, is common throughout the world. NDE reports of souls lost in a fog, as described by Howard Storm, are a warning.

Preparation for dying will one day become an integral part of ensuring the health and safety of individuals as they are released from their physical bodies. Obsession with the physical body and denial of the Life body is shown up as negligent ignorance in the light of after death evidence.

Embracing the possibility of continuous living could be the first step in alleviating unimaginable suffering, if NDEs, such as Howard Storm's and Ian McCormack's are taken seriously.

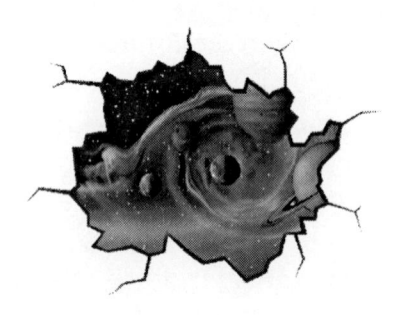

# Appendix 6

# Love in Action

Love in action is considering the welfare, of other people, other living things and the community we live in as preparation for continuous living in a situation of love, light and freedom. Love in action is parenting and partnering with love rather than anger. Love in action is putting compassionate love before passionate love. Love in action is mutual respect in the work place. It is volunteer and care work, nursing, visiting sick or old folk, trees planting, rescue and care for animals and picking up litter. Love in action is serving the welfare of the planet and society and living things. Love in action is selfless service; it can be in music and art, poetry and prose and anything that uplifts and expresses love and gratitude for Life as preparation for continuous living in a situation of love, light and freedom. Relationships can be a major test for love. We say 'I love you' but when the going gets rough, where is the 'I love you' then? Do we maintain the 'I love you' after parting or divorce or does in turn into 'I hate you'!

We think down on the poor and the disadvantaged and up to the rich and famous. The poor are our greatest asset because they provide an opportunity for us to give and perform acts of kindness. There are people who really struggle to survive on low wages and high rents.

If you are not struggling financially why not consider sponsoring a family that are? You could do things to help like finding a week's rent for them if they are in arrears or helping them buy food to feed their kids if they run short. Adopt a family or a single parent and give them **shelter if they are homeless** or money that you would spend on luxury items yourself. As a precaution ensure first that your giving does not support an addictive situation i.e. drinking, smoking, gambling or drugs.

# Appendix 7

# Helping Others Who are Earthbound

We can use the photo below to call on the assistance of Yeshua.

This photo was taken by a Dr Steinbeck on June 1st 1961. He was one of thirty archeologists working on a site at Chichen Itza in the Yucatan, Mexico. He was taking photographs when Yeshua suddenly appeared in a visible body and invited him to take a photograph.

If you are skeptic but concerned for yourself and find it impossible to pray or meditate, and are definitely not into anything else I have suggested so far, place a photo of yourself face down onto the photo of Yeshua. Close the book and put it in

## Continuous Living

your bookshelf. That action of intent will be sufficient to establish a connection with him.

We can help people who have passed over who we think may have become earthbound and lost in the fog. If you know of someone who has passed away you can help them into the light. You will need a photo of the person you intend to help. First look at the picture of Yeshua and ask him to help your friend then take their photograph and place it in this book face to face on the photograph of Yeshua.

Close the book and leave it for the magic to work. It is science actually. Photographs carry frequencies. They are a light imprint of the subject. By placing the two photographs together you are setting the intent to connect the deceased friend or family member to Yeshua. You do not have to be a believer in Yeshua. You can do this just for the deceased person you care for. Doing this exercise from a position of disbelief makes it even more powerful because the act is more unconditional and pure as an exercise of free will.

In the ancient wisdom teachings it is said that we on Earth can do more to help people who have passed over from the Earth than from any other position in the Universe. If you get into this you can do it with pictures of anyone who has passed away and Yeshua is the only one I know who can penetrate hell as he did for Howard and Ian and rescue people from the pit of despair. But they have to accept the rescue and welcome the light. All you can do is initiate the process and leave Yeshua to do the rest.

You can also use this exercise to help people you are concerned for who are still alive. If you are anxious about someone, they may be ill or in trouble or dying, place their picture in this book face to face on top of Yeshua and ask him to help them. It is so much easier then getting down on your knees and saying heaps of prayers like Sister Dolorous did for Howard Storm. When Howard spoke to her after his descent into death her only comment was "I wonder why it took so long?"[1]

1    **Storm H,** *My Descent into Death,* Clairview, 2000

# Appendix 7 – Helping Others Who are Earthbound

Don't expect immediate miracles. In fact don't have any expectations because there are other factors. People have a free will that cannot be overridden and they may have chosen a path of suffering or the way of darkness in hell as and essential part of their soul evolution. All outcomes are actually for you. It is our participation in the unconditional love and upliftment of humanity that matters.

If you don't have a photograph, conjure up an image in your mind of the person departed, focus on the picture of Yeshua and ask him to help them.

You can visualise a train of light to help people. Imagine you are on the platform with the person you are helping and visualise yourself leading them onto an imaginary train translucent with light that has just pulled into the station.

Play with your imagination. You could imagine the carriages studded with diamonds and gems sparkling with light. Or you could visualise them made of crystal. See the radiant smile on the face of the person you have helped as they step onto the train. Feel into their joy as they settle into a seat and are served champagne by angelic beings of light. You could imagine cherubs peering at them from over the seats in front, as playful children do on normal trains.

The play with imagination is important to engage your mind in the exercise when it is shouting 'oh how camp' or 'this is stupid' and other obnoxious kill-fun stuff. Counter negativity by really going to town in your imagination.

Paint a picture in your mind of Yeshua is in his overalls ready to drive the magnificent steam engine puffing in front. It is made of black obsidian lined with gold. There is no hurry to leave because Francis, who is supposed to be firing up the engine with shovels of coal, is distracted by a flock of birds that have landed on his head and shoulders and outstretched hands.

Yeshua is grinning as he leans out of the engine window with an engine driver cap on his head, watching as you pick up another lost soul. See yourself helping someone else you can think of off the platform and onto the train. Enjoy their delight as you take them into the restaurant carriage with white table cloths, laid with finest china and Victorian table lamps.

## Continuous Living

See the joy in their face as Mariam and Mary Magdalene dressed as waitresses offer them a menu and serve them food. As you leave the train they flash you a smile. Watch your passengers wave you goodbye as the train of living light pulls away. See their depth of gratitude for being helped on their way into the liquid light of eternal love.

Imagination play is the best way to pray. You can use this sort of play with your imagination to help people who are alive, as well as those who have died; visualizing them moving off in the train from darkness and confusion on the cold foggy platform into light and love in a warm carriage.

Using your imagination in this way is very powerful because it is a way you can take control of your mind, through the action of your will to make it work with you and for you rather than against you. The more you work your mind in this way, filling it with the images of the light, the more uncomfortable it will become for the usual resident negative thoughts that prefer to linger in misery, self pity and darkness; because you, by the power of your imagination, are transforming your mind into a space of living light where no darkness can dwell.

By helping others into the light in any way, you not only secure your own connection and safe transition into the light but could catapult yourself directly into the super-physical level of the Universe when your time comes to move on into the light. If you are acting like one of the masters of light, leading people that are lost from darkness into light, you may find yourself ascended as one amongst them.

There is no difference between any of us and any of them. It is compassion for others and acting selflessly out of love, in the living light of Life that leads to recognition and reception by the super-physical. The crew members of the crystal train were people like you and me. They had the same challenges and concerns, the same doubts and difficulties as we have. We are all one; we are all on the same journey from darkness into the light; from fear into love.

Be ready for all sorts of similar scenarios to pop into your head once you get started with this work of redemption (excuse the relingo). What I love about this book is I will sit down to write

## Appendix 7 – Helping Others Who are Earthbound

and it writes itself on a totally different subject to what I intended. Visualizing the train of light happened quite unintentionally.

In my mind I could really see Yeshua grinning from the cab, like a character from a cartoon, as the story unfolded. He is fun when you get to know him. If you can let go of your doubts and reservations and let go of being an adult to be as a child again, you will find you are just at the beginning a great adventure.

Yeshua said we have to be as little children to enter the Kingdom of Heaven. Think of him as an older brother. Turn to him if you are anxious. Look into his eyes in the photo and ask him to guide you on your way. You have nothing to fear when Yeshua is near.

# Appendix 8

# Happiness

An attitude of gratitude brings more joy to Life than grumbling and complaining. Living with love brings joy into living. Living with love replaces guilt, blame and shame with happiness. There is no need for concern about the future life if we live the life we have now with love in our hearts. If we live with love, happiness can unfold like a flower.

It is not being loved but loving others that brings happiness. Love for money and material things and being in love may bring happiness for a while but it doesn't last. Lasting happiness comes not from possessions or possessing people but from living in love with Life; the life that is always in us.

Love and joy can flow out from the Life within us, toward people, places and things, if we allow it. The objects of our affections continually change and sadness can come with the changes until we learn that happiness comes not from anything outside of us, happiness resides within us. That is why we need to know ourselves.

When we love whoever we are with a fountain of happiness opens within. For me recently it was a hornet that wanted to share a caravan with me. She was gorgeous and I was much happier when I welcomed her and let her fly in and out as she wanted than when I was in fear I might have to live in a hornets nest!

My son Sam's partner, Sonia, said that the best way to find happiness is to serve others. It matters not if we are rich or poor. Greed or envy will not make us happy but compassion, gratitude and service, even to a hornet, can release the flow of happiness. Howard Storm was taught that the best way to grow spiritually is through service. Many great spiritual teachers, including the late Saytha Sai Baba taught that. My experience is that this is true.

## Appendix 8 – Happiness

Life is generous and generosity encourages happiness to grow. Generosity and happiness are part of the same flow and poor people can often be more generous than rich.

The secret to happiness is doing the best we can to love and serve others. If we do the best we can as we are and with what we have, Life will provide us with what we need at the time when we need it. It is a mistake to think we need to accumulate things or we have to be successful to be happy. The really happy people are those that succeed with Love.

When I was married to Anna we bought a two bedroom house. It was a bit small as we had three toddlers and a baby. We had one bedroom and the children shared the other but within a month we had invited a homeless family with four littlies to share our home until they got sorted. We made room for our children on the floor of our bedroom so the family could sleep in the other room. I look back on that time as one of real happiness. I don't know how it was for Anna but she must have been happy too because she allowed another homeless couple to move into the living room with their Irish Wolfhound. The dog slept in the hall and I recall the little children climbing up and over it as they came and went from the house.

Today in *The Times* Bob Geldof is declaring his intent to open his homes to Syrian refugees and thousands of other big hearted people are doing the same. They will be rewarded with happiness for their generosity of spirit and demonstrate to the rest of us that the plight of refugees and the homeless is an opportunity not a burden. As we open our hearts and our homes to the dispossessed so Life will welcome us home when we are dispossessed of our physical bodies.

# Appendix 9

# Care for the Body

It is important that we take care of the physical body and do all in our power to maintain physical as well as mental health. To support this I have designed fortified superfoods to make sure we get all the essential nutrients we need in a natural form every day. I am a nutritionist as well as a physicist and I have many years of knowledge and experience in the field of nutrition.

As well as good nutrition it is important to exercise regularly but don't overdo it. Strenuous exercise generates free radicals that are accumulative in effect and do long term damage to the health. If you are involved in sport or vigorous exercise my superfoods can help as they contain powerful natural antioxidants to help quench free radicals.

I don't believe restricted diets are healthy. I recommend we should eat what we are offered with gratitude and include a little of all food types in our diets. As the Dalai Lama suggested 'Take meat for medicine' and the best meat for health is organ meat, most especially lambs liver. If you are concerned at the karma of eating meat, pray over it first to clear all fear from it and then send the animal love in retrospect.

Experts say we don't need supplementary vitamins and minerals if we eat a good diet but what is a good diet? A good diet is mixed and low in sugar, salt and alcohol but high in vegetables and rich in vitamins, minerals and trace elements . If you are young and healthy and your food is whole, organic and includes meat, fish and dairy products with liver twice a month then dietary supplements should not be necessary.

Most important of all our bodies need our love to function to the optimum. Love is the best medicine of all.

# Appendix 10

# The Love Revolution

Russell Brand is heading a Love Revolution today as Yeshua was heading a Love Revolution in his day and Brand's message is the same. We don't have to be religious; we just need to love Life and care for one another as we care for ourselves. Love is a word used for many things but the Love that Brand and Yeshua talk about is doing to others as we would have things done to ourselves.

From the vortex theory it is clear that while we are many, the consciousness underlying us all is one. We are all the one being in the many bodies. When we help someone we are helping ourselves. If we hurt somebody we are hurting ourselves.

Prayer is important – especially if we lose something or need a parking space – but prayer for others is the prayer that really matters. Look at what Sister Dolorous achieved by praying for Howard Storm for thirteen years. It is only thanks to her that I have written this book!

Meditation is what we do for ourselves. Russell Brand is encouraging clean living and the practice of meditation. Rather than asking for anything, meditation is a discipline of taking time to still the mind and experience the essence of Life within. Life Consciousness within each of us is who we really are. Through meditation we can know the I AM presence inside. Meditation is the practice of *Be still and know that I AM Life*.

My daughter Ondine is supporting the Love Revolution by maintaining a positive attitude to Life and serving her neighbourhood. She has organised the transformation of a neglected place in her street in London into a community space. She says lots of things happen to us in the day. If we focus on the good things and express gratitude for them we will be happy. If we focus on the bad things and complain about them we will be unhappy. Her secret for positive living is to establish positive habits of thinking. I am fortunate to have such wise children.

# Appendix 11

# Self condemnation

I am certain that people who have done bad things are not condemned by the things they have done because so much crime and bad behaviour is genetic or environmental in origin, there can't be judgment. Hell is more for the proud and the selfish than the bad! Only the cruel mind judges! I believe people judge and condemn themselves. That is a standard trick of the mind. First is tempts us into doing something then it beats us up with guilt, shame and blame for doing it. Finally it condemns us to hell for it. If we believe our thoughts then, by the exercise of our free will we condemn ourselves. That is why, as my daughter Josephine said, it is important that we learn to love and forgive ourselves for what we do and everything we have ever done.

If you have done something wrong, even if it is utterly unforgivable stand in front of a mirror, look into your own eyes – that is the hardest bit – and repeat three times:

*I am sorry,*

*Please forgive me,*

*Thank you,*

*I love you.*

If you have wronged someone, picture them in your mind and repeat those words to them or do it to a photo of them or ring them up and say the words over the phone or text them or tell them in person. These words can help us to release self recrimination, guilt and shame. Think of someone who has wronged you and say these words to them in your mind and see how your anx toward them evaporates. These words directed to parents can heal dysfunctional relations with them; especially if we feel they are in the wrong! As we forgive others it is easier to forgive ourselves. As we forgive ourselves it is easier to forgive others.

## Appendix 11 – Self condemnation

Calling on a guide like Yeshua to forgive us can help because he gives us the confidence to forgive ourselves. Even if Yeshua doesn't exist it doesn't matter. It is the principle of love and forgiveness he represents that matters. That principle is in our own hearts if we allow it to emerge despite the protests of our head that we don't deserve it.

Even if we doubt his existence, by allowing him to love us we give ourselves confidence to love ourselves. By loving and forgiving others as he recommended, we make it easier to love and forgive ourselves. His prayer, *forgive us as we forgive others* is teaching us how we can practice our purpose of breaking free of the slavery to our minds.

# Appendix 12

# Money

There is nothing wrong with money. It is what we do with it that counts. Money can provide an opportunity to practice love by giving. Many celebrities and the super rich do a lot of good by giving their money. A friend of mine was given a very expensive specialist wheel chair for her disabled daughter by the Beckhams. When it was delivered she was told by the driver he had delivered over a dozen all paid for by the Beckhams.

We hear the gossip and the bad news in the media while the good that many rich and famous people do goes largely unnoticed. Most people who despise the rich do so out of envy and if they have money the envious types don't usually give it away. Bill Gates said that making money is easy; it is giving it away that is the real hard work.

We don't have to have a lot of money to be generous; we can give a little bit from whatever we have. My experience is that the more I give, the more comes back to me; not just in money but in other things. My technique is to sponsor just one person working to uplift humanity. The individual shouldn't be related nor should we be in a relationship with them but they should be on the spiritual path so it is spiritual work we are supporting. I practice this *target giving* for me because of the joy and happiness I get from giving and to balance my own karma.

I used to worry about money but when I started to give it I stopped worrying about it. It is pointless worrying about getting money if you give it away. The amazing thing is it keeps flowing in because nature abhors a vacuum. I get anxious if I just focus on myself and my own needs. It takes trust to give when we really can't afford it but it is amazing how it comes back from someone else, or in some other way. Though I subsist on a small pension, I am never short of cash and I am loved and cared for wherever I go.

# Appendix 12 – Money

If everyone was giving, loving and caring for everyone else there would be no need for money and the Earth would be paradise. As James Redfield says in the 'Celestine Prophecy'[1], *"The world will only turn from darkness to light if we give money to people who inspire us spiritually."*

All too often spiritual people are hampered by lack of money. It is important to support people who are working to bring more love and light into the world.

We cannot be both right and have love. If we are right someone else is wrong and that sows the seeds of discontent, conflict and war. Humility is hard. Self righteousness is easy and criticism even moreso! Being right breeds anger and anger leads to the dark polarity of hatred. People are fixated with truth and righteousness but nobody can ever be certain of the truth and none of us know enough about anyone or anything to judge what is right or what is wrong. Being right is pride. Being an embodiment of love is to rise above the polarity of right and wrong and care for everyone. Money is the easiest tool for giving so before we judge what others are doing with money let us first consider what we are doing with it!

1    **Redfield** James, *The Celestine Prophesy* Bantam Books

# Appendix 13

# The Universal Law of Love

In the *Vortex Theory*[1] I put forward the idea that each vortex of energy is a space-time bubble that exists as a particle of motion relative to the space and time set up by another vortex. I call this the *Universal Law of Love* because every particle of energy depends on every other particle for its existence. If you are a particle of nothing but movement you must have space and time to move in or you could not exist. In the absence of absolute space and time you would depend on your neighbouring vortex of energy for your space and time and it would depend on you.

In *The Vortex Theory* I use the Universal Law of Love to explain Einstein's *Twin Paradox Theorem* for time. The Universe only works if every quantum is there for every other quantum. Love is fundamental to quantum reality. Love is quantum theory and relativity theory rolled into one. The whole Universe runs on *love thy neighbour!*

Ignorant of the universal law of love we have a history of going out to destroy our neighbour rather than love our neighbour and in the process we have almost destroyed ourselves. Nature is our neighbour and nature created the space for us to live in. In return we have crushed and exploited nature to the extent the Earth is now dying. Native people know and practice the universal law of love. They are custodians of nature with a deep respect for Life. Maybe the time has come for us to be humble and learn from them.

1    **Ash, D.** *The Vortex Theory* (Kima Global 2015)

# Appendix 14

# The Native elders

Elders from native peoples are now traveling throughout the world teaching their way which is to honour the land and nature, to honour the rites of passage from childhood into adulthood and the passage from this world into the next, and to honour the ancestors. Native people believe in spirit and continuous living and they know that our ancestors, those that have gone before us into spirit, can help us on our journey, when our time comes to leave the Earth.

I attended a gathering of native elders from Australia, Africa and South America. The native elders teach that the three most important things are to love, to forgive and to accept. One from the Amazon said, *"The mind is fragile, it can fall in any way unless we follow the heart."*

We went out from Europe with materialistic and monotheistic superiority. In ignorance of the people we met and with utmost savagery, we treated them as ignorant savages! Now we can learn from them about love, forgiveness and acceptance. They can teach us how to love Life and respect the Earth. They can teach us the value of ceremony and prayer and rites of passage and the value of simple living. They can show us how to connect with nature and teach us about humility!

With our scientific sensibility and our technological urban sophistication, with our university education and our city life dwelling we have lost touch with nature and the simple, spiritual values of the native peoples we crushed. So where are we now with our worship of money and our belief in death? Where are we going with our religion and materialism? We may know more but do we have more wisdom than the native elders with their feathers, pipes and drums?

# Appendix 15

# The Goddess

Two sacred laws of the Native Americans are *all are born of woman* and *never harm the children*. The malpractices of dishonoring, exploiting and abusing women and children are hallmarks of materialism and monotheism. In cultures where the goddess is honoured women and children are treated with respect. In our own society as women come into equality with men, they can also come into their personal power.

Every woman in her power is a goddess. Women tend to be more connected with the Life body than men. Wise are those who heed the intuition of a woman in tune with her Life-body. As women come more in touch with their inner goddess they are drawn to connect with other goddesses outside of themselves. Men also are drawn to the goddess.

In my opinion the premier goddess on Earth today is Amma. She has devoted her life to hugging humanity and she is also lifting India out of poverty by directing donations to building homes and hospitals, schools, universities and orphanages. She feeds the hungry, houses the homeless and consoles the distressed. She is mother to us all.

I am also devoted to Mariam the maiden mother of Yeshua and Quan Yin the female Bodhisattva. She was a princess, who like Buddha, left her palace out of compassion for humanity. I burn a candle to them both to balance the East and the West and maintain my multifaith practice.

We men need not be intimidated by the rise of the goddess. Instead of confronting a woman and attempting to overpower her we can be behind her to support her and help her come into her power. Then as a goddess she can turn round and receive us, and then as an equal she can help us men become gods.

# Appendix 16

# The Peace Education Program

Prem Rawat has a new initiative called *The Peace Education Program. (PEP)*. The purpose of this program is to help us discover our own inner resources; innate tools for living such as inner strength, choice and hope and the possibility of personal peace.

PEP is a non-religious and non-sectarian innovative educational program facilitated by volunteers. The curriculum consists of ten videos each focusing on a particular theme based on excerpts from Prem Rawat's international talks. The themes are: Peace, Appreciation, Inner Strength, Self-Awareness, Clarity, Understanding, Dignity, Choice, Hope and Contentment.

If you were to ask me who I think one of the wisest people on Earth I would reply Prem Rawat. I have listened to his discourses for many years and have always found them uplifting and inspiring. He speaks from the heart about matters that concern us all; issues that have nothing to do with our beliefs of disbeliefs but everything to do with making the best of ourselves and achieving the most out of our lives.

What I am endeavouring to say is that if you have a moment to go online now and can spare six minutes to listen to an introduction to the Peace Education Program, you might find it life changing and if there isn't a PEP in your area you can even apply to set one up.

Ultimately it is a matter of priority. Does Life really matter? Is it worth our while to attend a program of teaching on personal peace by one of the greatest Life teachers alive today? That is our choice. It could also be our greatest opportunity. Whatever you choose to do may Life bless you!

# Appendix 17: Soul Fragments

In his book Dispelling Wetiko Paul Levy wrote, *"Whichever name we use, we are in the midst of a collective psychosis of titanic proportions and one of its most stunning features is that very few people are even talking about it. Does that seem as crazy to you as it does to me? Our madness has weirdly become normalized to the point where we don't even notice it. This book is an effort to shine a revealing light on both the madness and the source of the madness which is ourselves."*[1]

In the wisdom of India the collective psychosis of humanity is perceived as a feature of the mind; not the Universal mind but the individualisation of mind, the psyche otherwise known as the soul. I believe the same psyche reincarnates again and again and the surviving Life-body from each incarnation carries a fragment of the psyche to its destination in the light or in the dark.

I think this accounts for what have been described as soul fragments. It may be the thoughts, inclinations and attitudes emanating from soul fragments in the extra-physical could influence the mindset of current incarnations in the physical. This is because each new incarnation is a soul fragment sharing the same individualised mind with all other soul fragments that have gone before it in generations of previous incarnations.

The persistent intervention of negative soul fragments in our minds could be the cause the collective psychosis identified by Paul Levy as Wetiko. Wetiko is a Native American name for the collective curse of evil that inflicts humanity like a psychic virus. If we are all influenced by our previous soul fragments, any residing in hell could account for psychosis. The number of soul fragments that have accumulated in either polarity could account for our karma. It is a matter of choice to what extent we follow the influences of our mind for good or for ill.

1     **Levy, P.** *Dispelling Wetiko: Breaking the Curse of Evil*, North Atlantic Books, 2013

# Appendix 17: Soul Fragments

This is only speculation but it makes sense of mental influences on people. I am now more inclined to believe it is the psychic fragments already in possession of the mind that influence people rather than extraneous psychic parasites. It makes sense of karma if people are influenced in this incarnation by their former incarnations.

The role of this karmic progress may be to act somewhat like an antagonistic muscle. It may be a process to build the strength of the spiritual muscle. The more a soul succumbs to evil, delivering soul fragments to the shadow lands, the stronger the antagonist that soul would have to work against in subsequent incarnations. Understanding this process we can appreciate evil is not something to be feared but something to overcome.

We could also come to understand the wisdom of accepting full responsibility for whatever predicament we find ourselves in as our progress in continuous living. There are no victims to tyranny only adversarial opportunities for soul growth. Armed with that understanding we can resist evil with good humour, loving the enemy as we fight it. Maybe that is why the Shepherd of Mankind advised us to love our enemies.

The theory of soul fragments sharing our minds suggests a person is not evil if they have evil thoughts or good if they have good thoughts. A person would be evil or good only if they act out the evil or good in their minds. We are all subjected to evil thoughts and good thoughts. Choosing to act out of goodness despite evil inclinations may engender growth in strength of spirit and purify the mind, enabling a positive feedback on a negative soul fragment. Maybe that is why the Shepherd of Mankind advised the whole law is in love.

While the majority may be inclined to act predominantly out of goodness, a minority choosing to follow the voice of evil in their heads may influence others. All too often they may rise to positions of power through ruthlessness. Psychopaths with this trait succeeding to positions of influence can give an impression of collective conspiracy to universal destruction.

Everyone has a degree of psychopathy due to negative thoughts – even if he or she refuses to admit to it. I believe evil influences exist in us all to test and tempt us. If the shadow self is not us now, but an influence in our minds from a past incarnation, it should not be a cause for guilt or shame. It can only destroy us if we feed it on a diet of hatred.

If we understand the shadow self, the darker aspect of our personality, as a soul fragment and treat it with love, through that love maybe we could redeem it. This idea of soul fragments is only a theory but if it works to shine light into the darker recesses of the mind then it may be worth considering. The value of this theory is it can give us a degree of detachment from our minds. The shadow, whatever its cause, gives us a better opportunity not to identify with our minds. Standing back from it we should be better able to choose between good and evil; to resist temptation and be true to ourselves.

Amongst the Yurok Native Americans being true to ourselves is achieved simply by giving our best to help others in need. Being true to ourselves is the one and only law of the Yurok. To love one another was also the one and only law for personal integrity given by Yeshua. This primary teaching of truth was given to Howard Storm by the Shepherd of Mankind when he was told all that is required of us is to love whoever we are with.

Teachers of meditation encourage us to divert attention from thoughts to make it more possible to choose a different direction and a fresh start. Because of the persistent nature of the mind it is necessary to maintain a consistent practice of distancing our attention from invasive thoughts for this to be effective. That is why the teacher, who can lead us from darkness to the light in our minds, is as important as the practice he or she recommends.

Without the teacher to encourage us to maintain the practice they recommend it can be impossibly hard to succeed against the machinations of mind. With the ongoing support of a teacher of truth to inspire us to meditate and live out of love it may be possible for us not only to save ourselves but to redeem our soul fragments through the mind we share so that together with them we can ascend.

When I speak of the Earth as the womb of angels, I believe there are three worlds associated with the Earth. The first world is the physical where good and evil coexist. The second and third worlds are in the extra-physical where the polarities are

## Appendix 17: Soul Fragments

separated. I believe a living plasmic being – colloquially know as an angel – seeds a new plasmic personality as a field individualisation of conscious mind in the womb of angels.

This is psyche or the human soul. I teach that the mind and emotions, representing the personality of that soul, reincarnate again and again through the generations leaving residual soul fragments as extra-physical bodies in the polarities of light or dark. There the soul learns the consequences of its actions or lack of action during its physical incarnation.

Because the mind of the soul is undivided, all the soul fragments have an influence on each other. In ascension the soul is eventually birthed from the womb as a new angel with all of its experiences in the light and in the dark serving as a wealth of wisdom for its future in service to the Universe. Some souls in the cycles of human incarnations are angels undergoing an upgrade. Some humans are angels or beings already ascended who come to Earth as teachers and instigators of change. They can come from the polarities of dark and light.

The shadow or the dark is not to be feared. It is an essential aspect of training for every evolving angel. I think the Earth, with its three worlds, one based on the speed of light and the other two based on twice the speed of light, could be the best training ground for gestating or upgrading angels yet devised in the Universe.

Credit has to go to Sanat Kumara – the Lord God of the three worlds, planetary Logos, the Ancient of Days, Gatekeeper of the Beyond the Beyond and Wolf of Mankind – for the outstanding success of his flight simulator for pilots of consciousness, as well as to his bother Sananda Kumara, the Shepherd of Mankind.

In their angelic forms as Lucifer and Michael, in the Judaic tradition, or the extra-terrestrial brothers Enlil and Enki in the Sumerian, these angel brothers hold the polarities of light and dark, good and evil, for the benefit of the spiritual evolution of all mankind.

# Appendix 18: The Dreamtime

The world view I would like to reference for the end of this book is the Dreamtime of the Aboriginal Australians. The Native People, who have inhabited Australia for fifty to seventy thousand years, believe there are three worlds namely the land, the human and the sacred. These correspond in our culture to body, mind and spirit.

The Native Australians believe that before the creation of the world and human life on Earth we existed in the Dreamtime. Now we live simultaneously in the world and Dreamtime. This idea fits with my concept of a Universal mind that individualises through incarnation and exists simultaneously in the extra-physical level of energy and in the physical. During sleep while the individualised mind is unconscious in physical time it is conscious in the dream time. At death the mind separates from the physical and returns to the Dreamtime.

The native people of Australia believe the Dreamtime incarnates in animals, plants and objects as well as in human beings. The ceremony of these enlightened people is to pray and sing to the Dreamtime – the mind – incarnate in the animal or object. They do this to connect with it and appeal to it; especially before they kill and eat an animal. This understanding is profound and simple. It makes sense in terms of the new understanding in quantum physics that all is mind and consciousness underlies everything.

Understanding we have a commonality and communion with everything, both animate and inanimate engenders respect for the Earth and all life upon her. The Yogis in ancient India discovered the vortex through communion of their consciousness with that of inanimate matter. Commonality between the Dreamtime understanding of the people of ancient Australia and quantum physics completes the circle between an ancient and modern appreciation of the Universe. We discover links between the most ancient and simple and the most modern and sophisticated. This is hardly surprising because the one thing we all share in common throughout all ages and on every continent is the mind.

# Bibliography

Alexander Eben. *Proof of Heaven,* Piatkus, 2012

Alexander Eben. *The Map of Heaven,* Piatkus, 2014

Allen D *When the Earth Nearly Died* Gateway Books 1995

Anglicized Bible *New Revised Standard Version* Collins 2011

Ash David & Hewitt Peter *The Vortex: Key to Future Science,* Gateway Books 1990

Ash David A., *The New Science of the Spirit,* College of Psychic Studies 1995

Ash David, *The Role of Evil in Human Evolution,* Kima Global Publishing, 2007

Ash David, *The Vortex Theory* Kima Global Publishing, 2015.

Ash Michael, *Health, Radiation and Healing,* Darton Longman & Todd, 1963

Banksy, *Wall and Piece,* Century, 2006

Berkson, William, *Fields of Force: World Views from Faraday to Einstein,* Rutledge & Kegan Paul 1974

Burr H.S, *Blueprint for Immortality,* Neville Spearman, 1972

Cagan Andrea *Peace Is Possible: The Life and Message of Prem Rawat* Mighty River Press 2006

Calder Nigel, *Key to the Universe: A Report on the New Physics* BBC Publications 1977

Clerk R.W. *Einstein: His Life & Times* Hodder & Stoughton 1973

Churton T, *The Gnostics,* Weidenfeld & Nicolson, 1987

Dawkins Richard, *The Blind Watchmaker,* Penguin, 1988

Dawkins Richard, *The God Delusion,* Bantam, 2006

Feynman R. *The Character of Physical Law,* Penguin 1992

Goswami A. *The Self-Aware Universe,* Putnam, 1995

Hawking Stephen, *A Brief History of Time* Bantam, 1988

Heley Mark, *The Everything Guide to 2012*, Adams Media. 2009

Hoyle Fred, *The Intelligent Universe*, Michael Joseph, 1983

Huxley A, *The Doors to Perception*, Penguin Books, 1959

Jeans James, *The Mysterious Universe*, Cambridge. University Press, 1930

Kuhn Thomas, *Black-Body Theory and the Quantum Discontinuity: 1894-1912* Clarendon Press, Oxford, 1978

Lewels J., *The God Hypothesis*, Wildflower Press, 1997

McTaggart Lynne, *The Field,* Harper & Collins, 2001

Nichols Preston, *The Montauk Project* Sky Books 1991

Perlmutter Saul., et al. *Discovery of Supernova Explosions...* Berkeley National Laboratory, Dec. 16, 1997

**Parnia** Sam (with Josh Young) *Erasing Death*, Harper One, 2013

Popa Radu, *Between Necessity and Probability: Searching for the Definition and Origin of Life*, 2004

Ramacharaka Yogi, *An Advanced Course in Yogi Philosophy and Oriental Occultism* 1904 (Cosimo facsimile 2007)

Rees Martin, *Our Final Century*, Arrow 2004

Redfield James, *The Celestine Prophesy* Bantam Books

Sabbah M & R, *Secrets of the Exodus*, Thorsons, 2002

Sartori P. *The Wisdom of Near-Death Experience*, Watkins, 2014

Sartori P. *The Near Death Experiences of Hospitalized Intensive Care Patients: A Five Year Clinical Study*, Edwin Meller, 2008

Sharkey, J., *Clinically Dead* Amazon 2012

Sheldrake R, *A New Science of Life*, Paladin Books, 1987

Storm Howard, *My Descent into Death*, Clairview Books, 2000

Velikovsky Immanuel., *Worlds in Collision*, V. Gollancz 1950

Walsch Neale Donald, *Tomorrows God*, Hodder & Stoughton, 2004

Wilhelm Richard, *I Ching*, Routledge & Kegan Paul, 1951

## Chapter 61 – Bibliography

For permission to quote extracts I would like to thank the following:

Alexander Eben. *Proof of Heaven,* Piatkus, 2012

Alexander Eben. *The Map of Heaven,* Piatkus, 2014

Banksy, *Wall and Piece,* Century, 2006

Berkson, William, *Fields of Force: World Views from Faraday to Einstein,* Rutledge & Kegan Paul 1974

Calder Nigel, *Key to the Universe: A Report on the New Physics* BBC Publications 1977

Clerk R.W. *Einstein: His Life & Times* Hodder & Stoughton 1973

Churton T, *The Gnostics,* Weidenfeld & Nicolson, 1987

Dawkins Richard, *The God Delusion,* Bantam, 2006

Hawking Stephen, *A Brief History of Time* Bantam, 1988

Heley Mark, *The Everything Guide to 2012,* Adams Media. 2009

Hoyle Fred, *The Intelligent Universe,* Michael Joseph, 1983

Jeans James, *The Mysterious Universe,* Cambridge. University Press, 1930

Lewels J., *The God Hypothesis,* Wildflower Press, 1997

Popa Radu, *Between Necessity and Probability: Searching for the Definition and Origin of Life,* 2004

Sabbah M & R, *Secrets of the Exodus,* Thorsons, 2002

Sharkey, J., *Clinically Dead* Amazon 2012

Storm Howard, *My Descent into Death,* Clairview Books, 2000

**Velikovsky Immanuel.,** *Worlds in Collision,* V. Gollancz 1950

Walsch Neale Donald, *Tomorrows God,* Hodder & Stoughton, 2004

Wilhelm Richard, *I Ching,* Routledge & Kegan Paul, 1951

# Index

**A**

Abraham 60

Akhenaten 62

Akhet-Aten 62

Alexander, Eben 9,13

Antigonus 139

Ascension 137

**B**

Bhagavata Purana 153

Blake, William 67

Brand, Russell 70,182

Burr, Harold Saxton 31

**C**

Cathar 67

Cathars 59

Chichen Itza 174

Chinese medicine 171

Churton, Tobias 66

Cold plasma 42

Constantine 65

**D**

DNA 34

Dawkins, Rfichard 126

Dawkins, Richard 56,58,61

Democritus 21

Descent into Death 17,38,131

**E**

Einstein 22 - 23

Emerson, Ralph Waldo 116

Enlil and Enki 149

Ermitage papyrus 152

Extra-physical reality 30

Ezour Vedam 155

**F**

Feynman, Richard 26 - 27

Freud, Sigmund 62

**G**

Global warming 157

Goddess 189

**H**

Hatshepsut Queen 152

Hawking, Stephen 13,26 - 27

Heaven and hell 15 - 16, 123, 142

Heley, Mark 43

Hell 54, 122

Hermes 167

Hermes trismegistus 48

Hermetic 22

Herodotus 151

Hesiod 153

Holographic universe 54,56

Hoyle, Sir Fred 117,146

# Index

Huxley, Aldous 34

**J**

Jesus Christ 58, 64

Jung, Carl 119

**L**

LSD 126

Law of speed subsets 29

Levy, Paul 195

Lewels, Dr. J 66

Life body 28, 31 - 32, 34

Life-body 41

**M**

Matrix 168

Maya 154 - 155

McCormack, Ian 28, 51, 70

Medicine wheel 170

Messodm Rabbi 62

Morgenstern, Christian 146

**N**

NDE 9, 11, 15, 19 - 20, 29, 31, 53, 126, 161

**P**

Papyrus Harris 152

Papyrus Ipuwer 152

Parnia, Sam 12, 16, 169

Peace education program 190

Planck, Max 9

plasma 42, 46

Pomponius Mela 151

Prayer 182

psychic phenomena 129

**Q**

Quantum Process 120

Quantum theory 22

Quark theory 27

**R**

relingo 8, 55

Role of Evil in Human Evolution 58

**S**

Sabbah, Roger 62

Sanat kumara 148

Sartori, Penny 8 - 9, 16

Satan 68, 148

Secrets of the Exodus 62

Senmut 152

Sheldrake, Rupert 117

Shiva 149

Storm, Howard 7, 14, 17, 23, 38, 70

super-physical 137

**T**

Tsytovich, V. N. 43

Tutankhamen 63

**V**

Valentius 66, 126

Velikovsky, Immanuel 151
Vortex theory 7, 11,13 - 14, 16, 22 - 23, 26, 32, 116, 131, 137, 182

**W**

Wetico 195

Worlds in Collision 151

Wyatt, Lucy 154

**Y**

Yeshua 138,176,184

Yeshua ben Yoseph 64

Yucatan 155

Yugas 163

**Z**

Zarathustra154

# About the Author

**David Ash** was born in Kent, England in 1948. One of eight children he was the second son Dr Michael and Joy Doreen Ash. David has been married three times and is father of nine children and grandfather of eleven.

David (with Peter Hewitt) is author of *The Vortex: Key to Future Science*, by Gateway books in 1990. He is also author of *The New Science of the Spirit* published in 1995 by the College of Psychic Studies in London.

In the same year Kima Global Publishing of Cape Town published his book *Activation for Ascension*. Kima Global Publishing has published all his subsequent titles including: *The New Physics of Consciousness* and *The Role of Evil in Human Evolution* in 2007 and *The Vortex Theory* in companion to this book in 2015.

David travels extensively speaking on the Vortex Theory and Continuous Living in a Living Universe. He is also a musician and singer-song writer. David usually includes his music in his teaching programmes and concludes his workshops with a Sacred Sabai Blessing which is a sung meditation based on a Thai style ceremony.

Lightning Source UK Ltd.
Milton Keynes UK
UKOW04f0242270216

269186UK00001B/6/P

9 781920 535773